COMPUTER SIMULATION MODELS

Other books of interest

G. U. YULE and M. G. KENDALL
An introduction to the theory of statistics

N. A. RAHMAN
A course in theoretical statistics
Exercises in probability and statistics (for mathematics undergraduates)

M. H. QUENOUILLE
Rapid statistical calculations

C-E-I-R LTD (SCIENTIFIC CONTROL SYSTEMS LTD)
Mathematical model building in economics and industry

"Griffin's Statistical Monographs and Courses"

A. Y. KHINTCHINE
Mathematical methods in the theory of queueing

P. WEGNER
An introduction to symbolic programming

R. C. GEARY and M. D. McCARTHY
Elements of linear programming: with economic applications

C. E. V. LESER
Econometric techniques and problems

COMPUTER SIMULATION MODELS

JOHN SMITH, M.A.

GRIFFIN  LONDON

Copyright © 1968

CHARLES GRIFFIN & COMPANY LIMITED

42 DRURY LANE, LONDON, W.C.2

All rights reserved

SBN 85264 100 1

First published in 1968

811,297

HA31.5
S

W7457435

Printed in Great Britain by J. W. Arrowsmith Ltd., Bristol 3

TO MIA

Preface

In recent years computer simulation has been widely used to describe and analyse the behaviour of complex dynamic systems of many types. But there are few general accounts of the subject available in book form. I have attempted to fill this gap by providing in one volume—

    (i) a grounding in the necessary parts of statistical theory;

    (ii) a description of the different types of model that can be constructed;

    (iii) an indication of application areas; and

    (iv) a discussion of practical aspects of model-building.

In order to illustrate the account I have included a description of several case studies in which models were actually constructed and run on the computer. The circumstances in which these studies were carried out were as follows. The study described in Chapter 9 was undertaken by the Operational Research Section of British European Airways, while I was working there in 1962. Chapters 10, 11 and 12 are based on three simulation studies carried out by the Mathematical Services Group of C-E-I-R Ltd (now Scientific Control Systems Ltd) on behalf of General Precision Systems (ATM) Ltd, the National Ports Council, and Stewarts & Lloyds Ltd, respectively. In all three cases the simulation work was part of a larger study undertaken by staff of the firms in question. I would therefore like to thank these firms for permission to describe the work and for many useful comments on my draft.

I would like to acknowledge the help of my brother who also made many useful suggestions.

I owe my greatest debt of gratitude to Dr M. G. Kendall. I would like to thank him for his interest, advice and encouragement at various stages in the preparation of the manuscript. Without him it is doubtful whether this book would have seen the light of day.

West Molesey,   JOHN SMITH
July 1968

Contents

Part 1: SIMULATION TECHNIQUES

Part 2: CASE STUDIES

PART 1
SIMULATION TECHNIQUES

Chapter 1

Introduction

It is, as a rule, difficult to analyse complex dynamic behaviour by means of mathematics. The equations that describe the behaviour of physical systems changing under the action of simple forces often prove to be intractable. The situation is even worse in the social sciences because here the system changes are the result of human decision-taking rather than of simple physical forces; this makes it difficult even to describe behaviour in terms of equations, and much more so to obtain satisfactory mathematical solutions.

As an alternative to mathematical analysis we can turn to numerical methods to solve these problems. We can proceed as follows. We can assume some initial state or condition for the system being studied, and we can then use whatever laws or rules of change we have in order to evaluate the states or positions through which the system moves as time advances through some stipulated period. By this means we can avoid altogether the use of mathematical analysis and the difficulties associated therewith.

Because this type of calculation proceeds in true time-sequence, we can regard it as "simulating" or copying the behaviour of the system under study. This becomes more obvious if the calculation is programmed for a computer. Then the machine appears to simulate the real-life system as its variables change; the machine looks as if it is mimicking changes taking place in the real-life system. We can say that we have a *model* of the real system, in the computer, which behaves very like the real system. By "model" in this context we refer not to an entity that physically resembles the real-life system but simply to a set of variables representing the principal features of the real-life system and a set of computer instructions representing the laws or decision rules that determine how these features are modified as time progresses.

3

With such a model it is possible to perform many trial runs to find the effect of alternative assumptions. The cost, inconvenience, and time involved in experimentation on the actual system itself, or on small-scale versions of it—the only alternative—can in this way be avoided. Indeed, in many cases such experimentation is not possible, due to political considerations or simply the irreversible nature of any change made in the real world. Therefore the use of a model usually offers the only practical means of obtaining accurate information on which to plan and design new systems.

Any system that can be described in logical or mathematical terms can be studied in this way. As much detail as desired can be included in the model. It is not necessary to make simplifications in order to obtain a solution, as is often the case with mathematical analysis. It is true of course that simplifications have to be made in order to conserve computer running time and to limit the time needed to formulate and check the model. Nevertheless, the ability to handle detail makes this type of approach very powerful.

We shall refer to this type of calculation, wherein system states are evaluated in realistic time-sequence, as a *simulation* of the system.

Random factors

Many systems of interest in real life cannot be described in completely deterministic terms: their future course of development is influenced by factors whose effect cannot be foreseen exactly. In many of these situations we can, however, describe the relationship between cause and effect in terms of probabilities. For each cause we can define a range of possible effects or outcomes and associate with each a certain level of probability. We can then simulate the effect caused by the action of a factor by sampling at random from the known probability distribution of its outcomes. When repeated many times in the course of a simulation run, this procedure correctly reproduces the average effect of the factor in the conditions assumed. From a purely statistical point of view the simulation calculation can then be regarded as a complex sampling exercise, the object of which is to combine certain given probability distributions, but where, because of the complex interactions between the model variables, it is

4

necessary to play out the events of the system in realistic time-sequence in order to obtain the correct combinatorial probabilities.

This, however, is a somewhat limited view of the purpose of a simulation calculation. On a broader view, simulation is seen as a tool for studying the dynamic properties of systems to gain some insight into how they work.

Fields of application

Due to the general nature of the approach, simulation can be used in a wide variety of fields of study. The following are the major fields in which simulation techniques have been used:

(i) *Engineering design*. Analogue computers have been used to design car shock-absorbers by simulating response to typical road conditions. Analogue simulation has also been used widely in the design of complex servo control systems.

(ii) *Econometrics*. Simulation models have been used to analyse and predict the behaviour of firms, industries, and national economic systems, and to study the effect of different control policies available to the authorities.

(iii) *Nuclear reactor design*. The flux of subatomic particles through reactor cores and shields has been determined by simulation methods, for use in reactor design calculations.

(iv) *Military operational research*. Simulation models have been used for the assessment of weapons and tactics under various assumptions about environmental factors such as terrain effects, weather, and other influences.

(v) *Management science*. Simulation has been used to solve problems of queueing and storage affecting the operations of individual firms and enterprises.

The models used vary considerably. Later chapters describe the types of model employed in each of the above fields.

To perform a simulation requires in general a large amount of computation, and it is only the advent of computers that has made the simulation method a practical means of problem-solving. While in principle it is possible to perform the calculations by hand, the time

5

taken is usually such that only one or two separate simulations can be achieved. With a computer, once the initial effort of model formulation and programming has been made, many "runs" can be performed at the expense of extra machine time only. Thus it is possible to study, relatively cheaply, how the system behaves under many different conditions. The knowledge gained can be regarded as synthetic experience of actual operations—experience which it is often impossible to acquire in any other way.

Most applications require a digital machine with a large data store because a comprehensive description of the system must be readily accessible at all times during the calculation. Large scientific computers are therefore the most suitable.

When applying simulation the following steps are taken. The first is to isolate, define, and quantify the important features of the system to be simulated. Then these features are represented in terms of a model which is programmed for the computer. The model is then tested out to make sure it is functioning correctly by making it reproduce known conditions. This validation stage enables the model-builder to check that all relevant factors have been included in the model and that the laws of behaviour have been correctly represented therein. The last stage consists of running the model to simulate the effects of various proposed system configurations, so as to assess relative costs and benefits.

Types of model

We can distinguish three main types of simulation model.

(1) *Continuous change models*

In this type of model the real-life system is represented by a set of continuous variables, changing continuously with time according to a set of differential equations. There are no stochastic effects. The time paths resulting are smooth curves.

Continuous change models are appropriate for representing the motions of physical systems acted on by Newtonian or classical electromechanical forces. They have also been used to represent activities of firms.

6

(2) *Fixed-period models*

In this type of model, time is broken up into a set of finite periods (usually quarterly or annual periods), and model variables are allowed to change at the end of periods only. The variables are continuous quantities. The manner in which they change is specified by a set of difference equations relating values at the current time-period to those at the immediately previous time-period. Very often stochastic or error terms are included in these difference equations to represent the fact that the variables are only somewhat loosely connected with one another in the real-life system because many minor, unpredictable influences also affect their inter-relationship.

Fixed-period models have been used mostly in Econometrics, where data are available only on a periodic basis.

(3) *Discrete event models*

In this type of model the variables are discrete quantities representing states of entities in the system. Interactions between entities take place at discrete points in time only, separated by intervals of inactivity. Such interactions are usually known as "events". Time is advanced usually from each event to the next. At each step all system changes implicit in current events are made, and any new events called in train are added to a future events list. The actual timing of events is usually affected by stochastic factors. Discrete event models are applicable to a wide range of problems concerned with systems of moving units. They have been used to model neutron flow in nuclear reactors, queueing and congestion of customers at service points, transport systems, and the interaction of combat units on the field of battle.

Continuous change models can conveniently be run on analogue computers, if high accuracy is not required. Fixed-period and discrete event models, on the other hand, dealing as they do with discrete changes separate in time, are more suited to the digital computer. There would seem to be considerable scope for models in which some changes are regarded as continuous and others as discrete, possibly run on hybrid computers, but little work has so far been done in this direction.

Chapter 2

Random sampling

In this chapter we discuss methods of representing unpredictable effects.

Unpredictability

In most of the situations we shall be concerned with in later chapters, we shall find some unpredictable elements. For instance, in a queueing situation, the arrival of customers cannot usually be foreseen exactly. In a combat between tanks the success or failure of a shot fired at a target cannot be accurately predicted.

Such unpredictableness is due to lack of knowledge of all the causal factors: we do not know the exact reasons for the timings of individual customer arrivals, nor the exact influences on the trajectory of the shell. Sometimes it is due to lack of a completely deterministic explanation of the phenomena in question, as in the case of the random walk of nuclear particles in a reactor. In this case we do not have a method of prediction for individual particles, even given information about initial conditions and the environment.

An unpredictable element appears to act in a random fashion, and we refer to such elements as *random* or *stochastic* elements. We do not imply by this terminology that the behaviour of these elements is intrinsically non-deterministic, but only that we cannot in practice determine the behaviour. Whether the fundamental laws of science are deterministic or not in the final analysis is a question we do not propose to enter into here. The point is that in many practical situations phenomena *appear* to be non-determinate, and we have to recognise this fact when simulating them.

Use of random numbers to represent stochastic behaviour

Suppose we have an event which can have three possible outcomes. As a simple example consider a tank firing a shell at a moving enemy tank. The result may be an outright miss, a hit inflicting damage only, or a "kill", putting the tank completely out of action. Suppose (as is likely in this particular case) that we cannot predict which of these three possible outcomes will happen in any given instance of the event. Then for purposes of simulation we can resort to a random selection technique such as coin-tossing to decide which one to choose. If we know that the probabilities associated with the three outcomes are, say, 0.2, 0.3, and 0.5, then we need a random selector which reproduces the outcomes, in the long run, with relative frequencies $0.2 : 0.3 : 0.5$. This can be achieved with a device generating random numbers between 0 and 9 with equal probability (as described in the next section), and a look-up table, as follows:

Table 2.1   Look-up table for outcomes

Random number drawn	Outcome
0, 1	Outcome 1
2, 3, 4	Outcome 2
5, 6, 7, 8, 9	Outcome 3

Generation of random numbers

We now discuss the problem of generating a sequence of random numbers, equiprobable over some range of values.

Many randomizing devices have of course been developed for use in gambling. Perhaps the most celebrated of these is the roulette wheel. A more sophisticated random selector is used in the Ernie premium bond machine. This relies on the unpredictable behaviour of electrons in neon discharge tubes.

However, these methods, good though they may be for games of chance, are not particularly suitable for use in computer simulation because the sequences of numbers they yield cannot be reproduced at any later time by the device.

We need reproducibility for two basic reasons. Firstly, when testing out a computer program the programmer needs to be able to use the same series of numbers over and over again so that he can see unambiguously the effects of the corrections he has made. The typical simulation program error will not show up until the model reaches some particular set of conditions. The programmer needs to be able to reproduce this set of conditions so that he can see whether his corrective action has been successful.

Secondly, when using a model to compare two systems, it is desirable to run the two systems under as closely similar conditions as possible. This means that the sampling values used should be identical in all runs (see Chapter 8).

We are thus faced with the problem of wanting to simulate random selection, while at the same time wishing to have repeatability in the sequences selected. We would like to have some deterministic method of generating a series of values that are sufficiently random *in appearance* to be usable for selecting samples. It turns out that we can produce a series of such *pseudo-random numbers* by applying a simple mathematical transformation over and over again to a suitably chosen starting value.

The transformation normally used is a multiplicative congruence of the form $x_{n+1} \equiv \lambda \cdot x_n \pmod{P}$, where x_{n+1} is the new value, x_n the old value, and λ and P are constants. In practice this means that each term is produced from the previous term by (1) multiplying by a suitable constant and (2) removing a certain number of digits from the leading end. In other words, multiplication and truncation are the only two operations required. It is clear that eventually the series will repeat itself. The length of the repeat period is k, where $\lambda^k \equiv 1 \pmod{P}$. If λ and P are suitably chosen, this period can be made extremely long, so that the possibility of repeats can be eliminated as far as practical sampling exercises are concerned. The method produces a very randomlike series. It is very convenient to program for a fixed word-length computer, since effectively only multiplication operations are required; and it has been chosen for all the simulation languages and packages described in the Appendix. The random numbers are usually normalized so that they all lie in the range (0, 1). If numbers are required uniformly distributed over any other range, say (0, N), they can easily be obtained by multiplying by a suitable constant (N).

10

Properties desirable in a pseudo-random number series

The series produced by a pseudo-random number generator should clearly possess randomlike characteristics. But what do we mean by "randomness"? Without entering into a lengthy discussion, we can say that for sampling purposes a series is adequately random if it possesses the following properties.

(1) *Uniform distribution*

The numbers should be uniformly distributed over their range. In other words there should be no "favoured" numbers, i.e. no numbers that occur more frequently than would be expected from normal chance variation.

To test for uniformity we take a large sample of the series (say 10,000 numbers). We divide the range up into x equal (non-over-lapping) subintervals. Then we see how many numbers fall in each of the subintervals. We know how many on average *should* fall in each subinterval (i.e. N/x, where N is the total number), so we can apply the chi-square test to see if deviations from the average are consistent with a random assumption. We can further refine this procedure by taking several sets of N numbers. We can then obtain a value of the chi-square statistic for each set, and inspect these values collectively to see if they are distributed as the chi-square statistic should be. We can form a distribution and inspect the value at various percentile points and apply the chi-square again to judge if this distribution is close to the expected distribution. We can also inspect the runs of pluses and minuses in the errors.

We may extend this test to two dimensions by studying the distribution of pairs of successive numbers over the two-dimensional existence space.

(2) *Independent terms*

An evenly distributed series may possess correlations among adjacent terms, which is clearly undesirable. We can test for such correlations by the Poker Test. Digits are generated and divided into consecutive groups of 2, 3, 4 or more. The frequency test is then applied to these groups. This test is known as the Poker Test because we can regard the groups as hands of cards at Poker, and we are looking principally at the frequency of rare combinations such as "3 of a kind". In particular, if the series of digits is divided into groups of 5 the following combinations can be recognised:

"Bust" (a b c d e); "Pair" (a a b c d); "3 of a kind" (a a a b c); "Full House" (a a a b b); "4 of a kind" (a a a a b); "5 of a kind" (a a a a a).

The frequency of these combinations can be compared with theoretical frequencies based on probability theory and a chi-square test may be applied to test consistency.

A series which passes both the test for uniformity and the Poker Test will be adequate for practical sampling purposes. However, many other tests of randomlikeness have been suggested. The reader is referred to references [1] and [4] for a discussion of these tests which are beyond the scope of the present volume.

Sampling from continuous distributions

We have seen how uniformly distributed pseudo-random numbers may be generated. Their use with discrete distributions, for example that describing the outcome of a shot from a tank, is quite straightforward, using the look-up table approach indicated previously. The use of sampling numbers to produce variates from continuous distributions is not so obvious and requires some further discussion. There are two main techniques that can be used:

(1) *The inverse transformation method*

Suppose the distribution we wish to sample from is described by a probability density function $f(x)$. We first form the cumulative distribution function $F(x)$—see Fig. 2.1. $F(x)$ varies over the range $(0, 1)$ as x varies over its range. For each sampling number in the range $(0, 1)$, say r_0, we thus have a point on the $F(x)$ axis as shown. This in turn defines a point on the x-axis, x_0. The following method is thus suggested: apply the inverse of the cumulative probability function as a transformation to random numbers uniformly distributed in the range $(0, 1)$. This method will produce the desired result because $F^{-1}(r)$ is a variable that has $f(x)$ as its probability

12

density function. This can be shown as follows. The probability that $F^{-1}(r)$ is less than or equal to x is

$$P\{F^{-1}(r) \leqslant x\} = P\{r \leqslant F(x)\} = F(x),$$

which is the probability that a variate with $f(x)$ as density function is less than or equal to x. The method, although simple in principle, is often difficult to apply in practice due to the difficulty of obtaining a mathematical expression for the inverse of the cumulative distribution function.

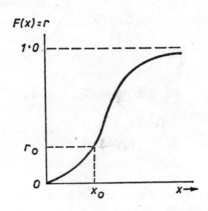

Fig. 2.1   **Cumulative distribution function**

(2) *The rejection method*

If the range of the variate x is finite, say (a, b), and $f(x)$ is bounded by an upper limit c, we can use the following method:

(i)   Generate a *pair* of random numbers r_1, r_2.
(ii) Use r_1 to locate a point y in the range (a, b) by the transformation $y = a + r_1 \cdot (b - a)$.
(iii) Use r_2 to locate a point z in the range $(0, c)$ by the transformation $z = c \cdot r_2$.
(iv) Reject the pair r_1, r_2 if $z > f(y)$ and return to step (i). Otherwise accept y as the value wanted.

Figure 2.2 illustrates the principle of the method. The pair r_1, r_2 defines a point P. The rejection criterion expresses algebraically

13

the condition that P is above the $f(x)$ curve at point $x = y$. If all such points are neglected, then the points that are left will be the points within the boundary of the curve. These clearly have $f(x)$ as density function.

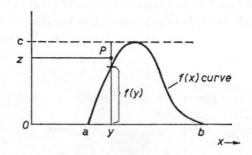

Fig. 2.2   Rejection method

References

MA 31·6 not on loan

[1] BIRGER JANSSON, *Random Number Generators*, Victor Pettersons, Stockholm, 1966.

[2] T. H. NAYLOR *et al.*, *Computer Simulation Techniques*, Wiley, 1966, Chapters 3 and 4.   T59·7   not on L

[3] K. D. TOCHER, *The Art of Simulation*, English Universities Press, 1963.

[4] S. GORENSTEIN, "Testing a Random Number Generator", *Comm. of Assoc. of Computing Machinery*, **10**, 2, Feb. 1967.

[5] E. D. CASHWELL and C. J. EVERETT, *A Practical Manual of the Monte Carlo Method for Random Walk Problems*, Pergamon, 1959.

Chapter 3

Analogue simulation of physical systems

The dynamic behaviour of physical systems, as we have seen, is often difficult to analyse by mathematical techniques alone. In such cases an alternative approach is to construct a computer model of the system which simulates its behaviour. This method has been frequently adopted by engineers when designing new systems, and has proved very useful.

The rules governing the behaviour of physical systems are derivable from the well-established laws of classical dynamics. When expressed mathematically they take the form of differential equations in system variables such as distances, masses, and forces. These equations completely define the time paths of the variables when starting values are specified. Hence the general problem confronting the design engineer is essentially that of finding the time-dependence of the variables involved, given a set of (ordinary) differential equations. The equations and therefore the problem are solved when all the variables are computed as functions of time.

Use of analogue computers

Unless a high degree of accuracy is needed, a convenient way of tackling this problem is to use an analogue computer. Analogue computers represent the variables and their derivatives by physical or electrical quantities (e.g. rotations or voltages) which are made to vary by the physical construction of the computer, according to the given mathematical relations.

Scaling of variables

To keep the variables within reasonable limits and to obtain maximum accuracy, the variables in a problem usually need to be scaled. The scaling factors used will depend on the range of variation expected. It is often necessary to scale the time variable too, so that the variables change at rates that are suitable for the components of the computer.

The time scale of the computer may thus be such that the computer variables vary faster or more slowly than the real-life variables. We may thus speak of "fast-time", "slow-time", or "real-time" simulation. The latter case is important because parts of the actual system being studied can be connected up to the computer, and their actual responses used in lieu of the computer representations. Another possibility is to use the model as a training device for human controllers. The model can be made to simulate the behaviour of a complex piece of machinery, such as an aircraft or nuclear reactor, so that trainee staff can learn to control the actual machinery by practising with the model. The model when used in this way is called a "simulator".

Electronic analogue computers

The most convenient type of analogue is the electronic type. It is possible to construct electric or electronic circuits to represent any set of differential equations by using certain basic computing elements. In this way a very large range of problems can be solved, in fact any problem that can be represented in the form of differential equations.

The basic element of the computer is the one that performs integration. There are two types: the integrations can be performed by either electrical or electronic means. Electrical integration is based on the use of capacitors, and electronic integration on the use of high gain D/C electronic amplifiers. Electronic integration is generally to be preferred because it is more accurate and practical.

The input parameters of the model take the form of the settings of variable resistors, and capacitors. The output is displayed by an ink pen writing on a rotating drum. The reader interested in the construction and characteristics of electronic analogue computers can find a good account of this subject in *Electronic Analog Computers* [1].

The general method of solving a problem by the computer is as follows. We first establish the equations of motion. We usually describe the system in terms of generalized co-ordinates and derive an equation of motion for each co-ordinate.

16

We then put together a specific block of computing elements to establish a "machine equation" corresponding to each of the given equations of motion. The output voltage from each block will represent displacements or velocities of the generalized co-ordinates. The input voltages to each of these blocks will represent (generalized) forces or accelerations. (Normally each such voltage is a function of one or more of the machine variables representing generalized displacements and/or velocities.) Electrical interconnections between blocks are then made to represent the coupling forces which logically connect the motions of the generalized co-ordinates.

Next we must decide on the initial conditions corresponding to the start of the run. We need to specify an output voltage for each of the integrators. These initial values are then set up in the machine and the run is commenced. An example is given below.

Electronic analogue computers have proved to be very useful because of the speed with which they produce the results of a computation, which is such that many different sets of conditions can be tried out in a short time. By watching how the response curve changes, the engineer obtains an insight into the working of the system he is studying. It is often this insight, rather than the actual numerical solution, which is important. The general form of the solution is usually what the engineer wants to know, and this is provided rapidly and pictorially by this type of computer.

It is possible also to construct *repetitive* electronic analogue computers. These develop solutions in a repetitive fashion, and display the result on a cathode-ray oscilloscope. The repetition rate is of the order of 10–60 cycles per second. This very rapid rate of solution enables the engineer to vary the input parameters, and literally watch the response change.

Illustrative example

To illustrate how a simulation run is set up on the computer we consider the calculation of a very simple ballistic trajectory, namely the flight path of a perfectly spherical cannon ball. Two external forces act on the cannon ball, gravity and air-resistance. This makes the problem sufficiently complex for a closed mathematical solution to be difficult. The object is to produce a set of range tables for the gun and also to investigate sensitivity of range to small changes in muzzle velocity and gun elevation. For simplicity we assume here that the aerodynamic force is proportional to the square of the instantaneous velocity of the cannon ball, and is a pure drag force (no force perpendicular to direction of flight).

Taking x and y axes to represent horizontal and vertical position respectively, we can write down the following two equations of motion for v, the instantaneous velocity, and θ, the angle of elevation of the flight path. For flight direction components:

$$m \frac{dv}{dt} = -mg \sin \theta - rv^2 \tag{1}$$

and for perpendicular components:

$$mv \frac{d\theta}{dt} = -mg \cos \theta. \tag{2}$$

We shall require the results in terms of x and y, so we need the two further relations

$$\frac{dx}{dt} = v \cos \theta, \qquad \frac{dy}{dt} = v \sin \theta. \tag{3}$$

We now have to construct four blocks of computing elements to represent the above four equations. These are shown in Figs. 3.1 to 3.4. The triangular box on the right of each diagram represents an integrating element. The square boxes represent multiplication by one of the varying voltages or a function thereof, and the circular boxes multiplication by a constant.

We now have to make the appropriate connections so that the outputs of the integrators are fed back as required. This is shown in Fig. 3.5 (page 20). The effect of scale changes is to modify the numerical constants of the equations. The actual circuit for computing the cannon-ball trajectory may as a result be slightly different from that shown in Fig. 3.5.

To obtain the solution corresponding to any particular set of starting conditions (i.e. muzzle velocity v_0 and elevation θ_0), we must now arrange for the machine variables to have the appropriate initial values at the commencement of the run. This means in practice arranging that the outputs of the integrators in the first two groups are set at v_0 and θ_0 respectively. Then a switch is thrown which makes the feedback connections linking the groups together, and the

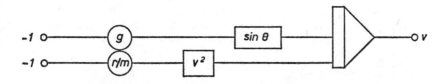

Fig. 3.1   Element block corresponding to first equation

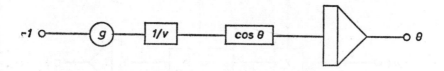

Fig. 3.2   Element block corresponding to second equation

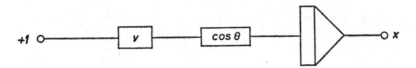

Fig. 3.3   Element block corresponding to third equation

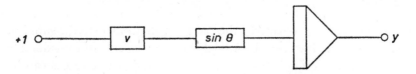

Fig. 3.4   Element block corresponding to fourth equation

computer variables begin to change in accordance with the stipulated differential equations, simulating the flight of the cannon ball through the air.

19

c

The type of solution obtained from such a computation is shown in Fig. 3.6 (opposite), in which the time paths of the four machine variables are plotted out.

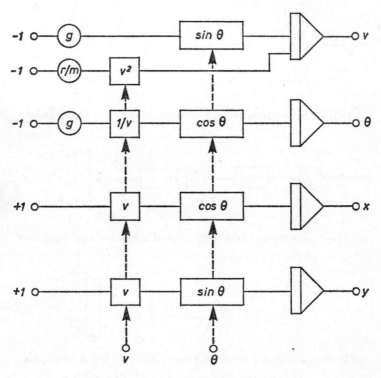

Fig. 3.5   Complete computing circuit

So far we have considered the cannon-ball problem in the simplest possible form. However, the model is clearly capable of considerable refinement if required. For instance, the aerodynamic forces can be treated more exactly, and also the effect of the earth's rotation can be allowed for by extra terms in the equations of motion. Such refinements call only for additional computing elements, and not for more sophisticated mathematical methods, which would be the case if a closed analytical solution was being sought. This is one of the inherent advantages of the simulation approach.

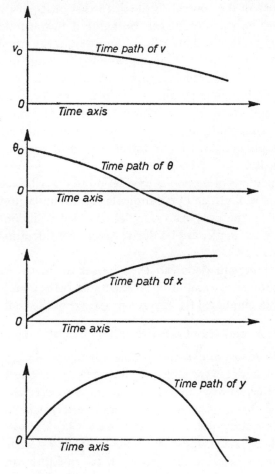

Fig. 3.6   Time paths of variables

APPLICATIONS

(1) Trajectory calculations

Rocket trajectories have been computed by simulation. The equations are complicated by the need to recognize that the rocket's mass is not constant, but diminishes as fuel is used up.

Aircraft flight manoeuvres have also been studied by simulation. The motion of an aircraft in the vertical plane has been derived for given settings of the control surfaces. In this problem the system of forces acting on the aircraft can be reduced to the following four: thrust, weight, drag, and lift. Thrust and weight can usually be taken as constant forces, the first acting along the longitudinal axis of the aircraft, and the second acting vertically downwards. The drag is a complex function of aircraft velocity, angle of attack (angle between longitudinal axis and direction of flight), and air density. The lift is a complex function of aircraft velocity, angle of attack, the rate of pitching of the aircraft, the control surface deflections, and air density.

Two equations of motion can be set up to describe acceleration in the direction of flight and perpendicular to it (similar to those of the cannon ball in the previous example). A third equation, describing changes in pitch angle, can be derived from consideration of angular momentum.

These three equations can then be set up in the computer by means of groups of computing elements as before, and the motion of the aircraft simulated for any given settings of the control surfaces.

(2) Vibration and shock-absorbers

A different sort of dynamic problem is that of designing a system to absorb or mitigate the effect of vibrations or shocks. In the trajectory problems just discussed the external forces were either constant, or slowly changing (namely gravity and aerodynamic forces). The resulting trajectory was a smooth curve. Shock-absorbers, on the other hand, are acted upon by rapidly varying external forces. The problem is to produce a design that smooths or damps out variations.

The external force may be a systematic vibration, represented by the equation

$$F(t) = a \sin \omega t,$$

or a sudden shock represented by a step or ramp function

$$F(t) = \begin{cases} 0 & (t = 0) \\ a \quad \text{or} \quad at & (t > 0). \end{cases}$$

The first type of force occurs in situations involving machines containing more or less unbalanced components rotating at a circular frequency ω. The second occurs in the design of car suspension systems, where it represents a sudden step or ramp change in the road surface.

The approach of the design engineer is to interpose an absorbing mass between the external force and the main body whose stability is required, and to couple the two masses with suitable springs and dashpots. A "spring" in this context is a device whose resistive force is proportional to its *compression*, and a "dashpot" one whose resistive force is proportional to its *rate of compression*. The general problem is to decide on the strengths of these damping components.

We can set up the equations of motion on the computer as before and solve for different values.

(3) Feedback control systems

Simulation is useful in determining the best way of controlling a complex dynamic situation. Some examples are given below. As before, the method is to represent the essential features of the dynamic system—in particular, response to control activity—and then reproduce its behaviour under prescribed sets of conditions and control rules.

A simple automatic control set-up is shown in Fig. 3.7. This can be regarded as a simple servo mechanism. The idea is to be able to control the movement of a heavy load. We have a motor with power supply connected up to the load, and a sensing device to sense the position of the load at any time. The sensing device produces an

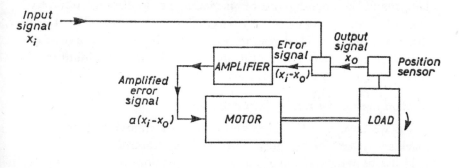

Fig. 3.7 Simple servo control

23

output signal, which can be compared with any input signal representing the desired position of the load. The difference between the two signals is fed back as an error or correcting signal, to control the movement of the motor, via an amplifier.

Given such a system, we wish to know how it performs; in other words, we wish to know the response of the servo to changes in the input signal. In particular we wish to know how the load moves under the influence of a step change in the input signal.

We can derive an equation of motion for the load as follows. Let the input and output signals be x_i and x_0 respectively. Then the feedback signal will be $a(x_i - x_0)$, where a is a constant. The driving force on the motor is thus $b(x_i - x_0)$, where b is another constant. Hence the equation of motion for the motor is as follows:

$$(I_L + I_M)\frac{d^2x_0}{dt^2} + r\frac{dx_0}{dt} = b(x_i - x_0),$$

where I_L and I_M are the moments of inertia of the load and the motor respectively, and r is the motor damping coefficient.

This equation can be used to explore the transient behaviour of the system to step changes. Values of the parameters can be chosen to prevent any tendency towards uncontrolled oscillations or "hunting".

Much more complicated examples of feedback, or closed loop, control systems can be equally well studied in this way.

Three typical responses to step changes are shown in Fig. 3.8. The first diagram shows a case of uncontrolled oscillation, with absence of any damping factor. The second diagram shows what happens at the opposite end of the scale—no oscillations, but very sluggish response due to very large damping factor. The third diagram shows the ideal response to a step input: the curve rises rapidly at first, then homes onto its final value with a minimum of oscillation.

Manual control: instruction by use of models

As we have noted, it is possible to use a model of a complex machine system for the purpose of training controllers. The main field has been the training of pilots with flight simulators. In addition to the advantage of cheapness, flight simulators can reproduce emergency situations without any element of risk; they are thus invaluable for teaching emergency recovery drills. Nuclear reactor simulators have also been constructed as a training aid for operators learning to control, start up, and shut down power reactors. The response to control actions (e.g. the moving of control rods) is complex,

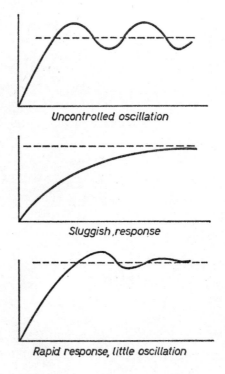

Uncontrolled oscillation

Sluggish, response

Rapid response, little oscillation

Fig. 3.8   Three types of response to a step input

due to the chain nature of nuclear reactions within the core; here, too, emergency recovery drills can be harmlessly taught. Simulators have also been used by driving schools.

It must be stressed, however, that these applications are distinct from the use of simulation as a means of studying a complex system in order to gain better understanding of its behaviour, and for this reason they are not discussed further in this book.

References

[1] G. A. KORN and T. M. KORN, *Electronic Analog Computers*, McGraw-Hill, 1956.

[2] S. EILON and D. P. DEZIEL, "The Use of an Analogue Computer in some Operations Research Problems", *Op. Res. Qly*, **16**, 3, Sept. 1965.

Chapter 4

Models of firms and industries

(1) Information feedback models

With the success of simulation models in the engineering control field, it is interesting to speculate whether similar models would be useful in the analysis of individual firms and industries.

In this connection J. W. Forrester [1] has shown that it is possible to represent the dynamic behaviour of an individual enterprise as an information feedback control system whose behaviour can be specified by a series of differential equations. He has further shown how these equations can be turned into a simulation model for a digital computer so as to enable various control policies to be tested out, with the object of determining a set of controls that smooth the fluctuations created by changes in demand patterns.

In the Forrester model the firm is represented by a network of continuous flows of six possible types of commodity, namely orders, material, money, personnel, capital, and information. Information plays a special role, because it is through information that control is exercised over other flows. The system is regarded as being controlled by information feedback. The rate of each flow component is assumed to be continuously monitored and compared with the target rate. Error signals are fed back via the information flow network to modify each flow so as to make it conform more closely to desired rates.

Although this type of model could be solved using analogue computers, as described in the previous chapter, it is better to use the digital computer because of its greater accuracy and the ease with which it can handle a large number of variables.

The time paths of the system variables are evaluated by progressing through time in discrete, equal-sized time steps. The differential equations of the model are replaced by corresponding finite difference equations which are solved at each time step. The time step used is sufficiently small for there to be effectively no change in rates of flow during it.

There are two basic types of variable, namely *levels*, which stand for physical stocks, financial balances, and other accumulations in the system; and *rates*, which stand for the rates of change or flows of these commodities from one point to another.

The values of all variables are computed at each time-interval from two main types of equations:

(a) *Level equations*

These determine the way in which levels change from one interval to the next. For example, a physical stock-level equation might have three terms representing the value of the stock level at the beginning of the solution interval, the rate of inflow during the interval, and the rate of outflow.

(b) *Rate equations*

These determine the way in which the rates change from one interval to the next. The effect of feedback control is represented via rate equations. It is usually necessary to have certain auxiliary and supplementary equations also, which simplify the main equations.

The model is now completely specified, apart from initial conditions, which are inserted via initial-value equations.

It is possible to insert terms into the equations to represent transit delays between any two points of a flow network. Fluctuations in the flow rate at the upstream point are then not transmitted instantaneously to the downstream point, but only make their effect felt there some time later.

It is not possible, however, to represent in this way queueing delays caused by insufficient capacity. Indeed, queueing and congestion effects cannot be accurately represented in this type of model. If these are significant it is necessary to use a model of the type dealt with in Chapter 6.

To assist in setting up the computer program there is a specially designed programming language called "Dynamo". For further information see [1].

Applications

There have been few applications of this type of model to actual firms, in spite of the obvious generality of the approach. The best-known example is the modelling of the Sprague Electric Company, an American manufacturer of small electrical components—see [1]. The reason why this type of model has not been applied more widely is twofold. Firstly, such a model tends to become very complex if all six flow networks are included and described in detail. Secondly, the model requires a large amount of data before it can be run. Much of the data is difficult to obtain: in particular, data relating to the information flow network. It is not easy to determine all the many channels of communication that exist within a firm.

For these reasons it has been found difficult to use this type of model to provide reliable numerical solutions to actual management problems. The approach is useful, however, as a means of providing a qualitative description of the behaviour of a firm under alternative methods of control.

(2) Econometric models

Simulation has been widely applied to the study of the dynamic behaviour of larger economic systems. Models of industries and complete national economies have been constructed. Some of these models have been used to study the validity of certain econometric theories, while others have been used to test the effectiveness of possible control policies.

In essence, the function of an econometric model is to trace out the time paths of a set of economic variables such as output, price, inventory, employment, and so on. These variables are regarded as continuous quantities. The model computes their values at fixed intervals of time, usually monthly or quarterly.

Certain variables are considered to be external to the economic system, in the sense that they act upon it but are not acted upon by it. These are termed *exogenous* or *autonomous* variables. The model can be thought of as a means of determining the response of the economic system to changes in the exogenous variables.

The model is defined by a set of equations expressing variables measured at time t in terms of variables measured at time $t - 1$. The coefficients in these equations are estimated from economic data. Methods of estimation, often complex, depend upon theoretical considerations beyond the scope of the present volume.

Random terms are often included in the equations to represent the action of the many minor influences whose effects are unpredictable. Random sampling methods as described in Chapter 2 are used to supply values of these terms at each time-period.

In the simplest models the equations can be solved independently, because the right-hand side of each equation contains only lagged values of variables, which are all available from computations already performed for the previous time-period. In more complex models there can be terms in non-lagged variables also, in which case the equations cannot be solved independently but have to be solved simultaneously. A model having such equations requires a very large amount of computing which is time-consuming, even for fast computers.

The construction of econometric simulation models is fraught with many difficulties which are fundamental and peculiar to the science itself. For example, there is the problem of which units to use for measuring such variables as price. There is a lack of data on many variables; also there is the difficulty of validation.

Models of an industry

A number of models have been constructed to study the behaviour of the price and production quantities of goods, under given assumptions about supply and demand. For instance, we may in this way study the consequences of the hypothesis that the quantity demanded of a certain commodity in any time-period depends primarily on the price prevailing in that period, and that the quantity supplied depends primarily on the price prevailing in the *previous* period.

The model shows that the price may behave in one of three ways: it may oscillate with ever-increasing amplitude; it may oscillate with constant amplitude; or the oscillations may be damped out within some finite period of time, to reach a stable position of equilibrium.

Various refinements and extra factors can be incorporated. The first of these is that random terms can be added to represent unpredictable factors affecting demand, supply, and the balance of the two achieved in the market. These random variables are represented by repeated sampling from distributions as described in Chapter 2. Secondly, an account of inventory holding can be incorporated. An assumption is made about the influence of inventory on price, such as: if inventories in the previous period fall, the price of the commodity rises, and vice versa.

So far we have not included any account of competition within the industry. We can do so by considering several individual firms operating within the industry, each making decisions about how much they will produce at each time step. We can assume that the object of these decisions is to obtain the maximum profit for each firm.

We can have a demand function for the industry as a whole, as before, with a random term to represent unpredictable effects. The price prevailing at each step can then be found by seeing at what price the market can absorb the total output from the industry.

These models are capable of considerable elaboration to make them realistic and applicable to particular industries at particular stages of development.

Models of a national economy

The behaviour of a national economy can also be studied. The number of variables involved can become very large, depending on the level of aggregation chosen. One important study in this field took as its subject the behaviour of the United States economy in a period of recession [5]. The object was to see how vulnerable the economy was in general to the onset of depressions, and in what way different automatic stabilizers would work to lessen the depth of the trough, and the elapsed time before recovery.

All variables were treated in a de-seasonalized way, to remove any seasonal swings that might obscure the underlying effects. The time-period taken was quarter-year. The main variables were GNP and total consumption. These were computed at each time step. The stimuli or external forces on the system in this case were taken to be (1) government purchases of goods and services, (2) gross private domestic fixed capital investments, and (3) foreign investments. The model computed the movements of consumption and

31

inventory investment in response to movements in these factors. Several experiments were made with this model.

Firstly, a medium recession was tried out, representative of recessions experienced since the Second World War. This was primarily a validation run. The external forcing function (autonomous demands on the economy) took a dip and then recovered.

Secondly, a more severe recession was tried out, the forcing function being a two-year period.

Thirdly, a step change or shock was tried out. In this last experiment the model was run both with and without an inventory factor. The results showed that inventory reactions tend to create an oscillatory response about the new equilibrium position. At first the GNP and consumption variables dip below their equilibrium values, and then recover; when the inventory factor is omitted, however, the response is well damped and no overshooting occurs. These situations can be compared with the reactions of physical systems under step changes to the stimulus, illustrated in Fig. 3.8.

In each of these experiments tests were made of various fixed policies as means of stabilizing the reaction. Two policies in particular were investigated. Firstly, increases in redundancy payments were inserted into the model. These appeared to have only a moderate effect. Secondly, rates of income tax were inserted as functions of the decline in GNP. Thus tax collections were supposed to stop completely when GNP fell by one-sixth, whereas a decline in GNP of one-twelfth was supposed to cut tax by only 50 per cent. The second policy proved to be a much more effective means of reducing the depth of the depression than the first.

References

[1] J. W. FORRESTER, *Industrial Dynamics*, Wiley, 1961.

[2] T. H. NAYLOR *et al.*, *Computer Simulation Techniques*, Wiley, 1966.

[3] R. M. CYERT, E. A. FEIGENBAUM and J. G. MARCH, "Models in a Behavioral Theory of the Firm", *Behavioral Science*, **4**, April 1959.

[4] J. V. YANCE, "A Model of Price Flexibility", *Amer. Economic Review*, June 1960.

[5] J. S. DUESENBERRY, O. ECKSTEIN and G. FROMM, "A Simulation of the United States Economy in Recession", *Econometrica* (1960), **28**, 4.

Chapter 5

Nuclear reactor calculations

In this and succeeding chapters we discuss models of a rather different type, which track the motions of individual particles, entities, or units in order to obtain information about the evolution of an assembly of such elements. We commence by considering the simplest of these models: that in which interactions between particles are ignored and particle tracks are simulated one at a time. This type of model has been much used in connection with nuclear reactor design. When designing nuclear reactors it is necessary to consider the flow of subatomic particles, principally neutrons and photons, within the reactor core and through the shielding. It is difficult to determine what this flow is for any design. One approach is to represent the flow as a continuous flux quantity, and use the basic law of continuity that must hold at every point inside the reactor and across the shielding—i.e. that the flow into every small volume must equal the flow out plus any flow created within the volume—in order to derive differential equations for the flux. When solved, these equations yield the flux pattern across the reactor core and through the shielding.

In fact, of course, the flow is not a continuous flux: rather it is composed of innumerable minute particles, flying hither and thither in all directions. An alternative approach is, therefore, to represent the actual motions of the particles. If we simulate a sufficiently large sample of typical particle histories, we can build up a picture of the flow in the reactor, and derive average flux properties. This approach can often give a more accurate answer because fewer restrictive assumptions have to be made.

The procedure is to take a given point in the reactor and assume that there is a source of particles there. By tracing a large number of particle paths we can then estimate what percentage of particles can be expected to pass beyond any stipulated physical boundary, such as the outside face of the radiation shield, or the percentage of particle paths having any other property of interest, such as leading to a nuclear fission.

Neutrons and photons carry no electric charge. They are thus unaffected by fields of force within the reactor, and their trajectories consist of a series of straight-line free flight paths between collisions, or discrete interactions, with the surrounding matter (see Fig. 5.1).

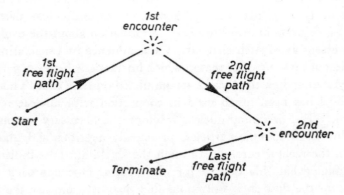

Fig. 5.1 Typical particle path in a reactor

Interactions between the particles themselves can be ignored as rare events. It is possible, therefore, to simulate the complete history of each particle on its own, without reference to any others.

The length of each free-flight path component depends on the density of the medium through which it is passing, the energy of the particle, and the types of atoms present. These factors do not, however, define a unique path-length in any given case as the atomic nuclei are not arranged regularly along the path. We must therefore assume that path-lengths are subject to some random variation.

We assume, in fact, that the atoms are distributed perfectly randomly along and in the neighbourhood of any given path. The probability of interaction while the particle is traversing a small distance δs along the path is then proportional to δs, say $\sigma \, \delta s$, where σ is known as the "interaction cross-section". σ depends of course on the energy of the particle and the nuclear properties of the reactor core material, as well as on the density of the material. It follows from the above assumption that path-lengths will be distributed according to the negative exponential distribution with mean $1/\sigma$. To reproduce free paths in the model we therefore sample from a negative exponential distribution with this mean.

At any given encounter, several types of interaction can occur, such as elastic scattering, inelastic scattering, fission, capture, etc. The laws governing these interactions are known only in terms of probabilities, so we cannot say which interaction will occur at any given encounter. However, we know the probabilities associated with each kind of interaction, so we can use random sampling to determine the type of interaction which takes place at each encounter.

It is now clear how the simulation calculation should proceed, i.e.:

(a) Choose a direction from the source.
(b) Pick a free path length by random sampling, using a negative exponential distribution with mean determined by particle energy and the type and density of material in that region of the reactor.
(c) Pick a type of interaction by random sampling from the interaction probabilities appropriate to the particle energy and material.
(d) Find an on-going path direction, again by sampling from distributions appropriate to the type of interaction (e.g. elastic scattering or inelastic scattering), and return to (b).

Eventually one of the conditions that we are looking for, such as escape from the shield, will occur, or the particle will be absorbed without achieving any of these conditions. In either case the calculation for that particle can be terminated and the result recorded. The next particle can then be taken.

When a large number of particles have thus been traced, the relative frequencies of the different possible terminating conditions set up for the problem can be calculated as required.

The two problems that have been most studied with this type of simulation, sometimes called a "random walk calculation", are (i) the shielding problem, and (ii) the criticality problem. In the

35

first it is desired to determine what thickness of shielding is necessary to protect operators against nuclear radiation emanating from the reactor. In the second it is desired to determine what size of reactor core is just "critical" in the sense of being just able to support a nuclear chain reaction.

Variance reduction techniques

The chief difficulty in these simulations is to calculate a sufficient number of particle histories to give an answer accurate enough for design purposes. There is considerable variability introduced by the sampling of free path lengths from negative exponential distributions. Answers may be required separately for different directions within the reactor, so that a reasonable sample is needed for each direction. For these reasons a simple simulation as described above is seldom capable of yielding sufficiently precise results within a reasonable computing time. Hence it is necessary to seek ways of improving the precision of the results without resorting to larger samples.

Among methods which can be used are (i) the technique of antithetic variates, and (ii) the use of control variates. These are described in Chapter 8.

It is interesting to note that the need to use these techniques is less pressing when we are modelling industrial, economic, or military situations, because in these it is usually possible to simulate a large slice of real time without incurring prohibitive computing costs; indeed it is often feasible to simulate the whole of a period of interest at fairly modest cost. In the nuclear reactor calculation, on the other hand, it is possible to track only a minute fraction of the millions of particles that exist at any time within the core. Hence sampling errors can be much more serious.

References

[1] E. D. CASHWELL and C. J. EVERETT, *A Practical Manual of the Monte Carlo Method for Random Walk Problems*, Pergamon, 1959.

[2] J. M. HAMMERSLEY and D. C. HANDSCOMB, *Monte Carlo Methods*, Methuen, 1964.

36

Chapter 6

Models of queueing and storage

In this chapter we consider the application of simulation to problems of queueing and storage. These arise in many situations both in service and manufacturing industries. It is convenient to consider these problems under six main headings.

(a) Customer service

It is often difficult to arrange to serve customers immediately on arrival, and some degree of customer queueing is often inevitable. Consider the simple examples of the service of customers in a post office or the checking out of shoppers at a supermarket. Because of the irregular arrival pattern of customers at the service points there are periods when many customers arrive, more than the servers can deal with immediately, with the consequence that some have to wait in a queue. Some customers may take unusually long to serve, in which case other customers may be held up. Many other examples from everyday life can be thought of where fixed service facilities have to deal with fluctuating demands. The irregularity of the demand pattern is usually due to the fact that customers are acting independently of each other. A good example is that of telephone subscribers demanding lines from a central switchboard. Sometimes the pattern is made even more peaky by the influence of external factors. For instance, ship arrivals at a port can be bunched by the action of tide or bad weather.

For Management the following problem arises. Given that the pattern of customer arrivals over a period is such and such, what service arrangements should be provided, bearing in mind that different service arrangements will force different degrees of queueing on the customers applying for service?

Except in conditions of monopoly, excessive queueing will cause customers to transfer their custom to competing service organizations. This fact acts as a considerable incentive to managements to contain queueing levels to those acceptable to customers. Often managements will aim to provide a better standard of service in terms of waiting times in order to win more business.

We can formulate the queueing problem therefore as follows. What service arrangements are necessary (or what is the best arrangement of servicing customers) in order to provide some predetermined standard of service (measured in terms of customer waiting time) set by Management?

Sometimes an actual money cost is associated with delays to customers. For instance, delays to airline passengers often involve extra meals or hotel expenses chargeable to the airline. Delays to ships using a port must often be paid for by the ship charterers in demurrage charges, if the days in port exceed a stipulated number. In such cases the queueing problem can be regarded as the problem of determining the facilities needed to provide an adequate standard of service at the least cost to the supplier of the service, taking into account both the cost of facilities and cost of delays.

In many cases the only practical solution is via simulation. The approach consists of playing out in the computer all the moves made by both customers and servers throughout a typical period of time, and collecting data on queue lengths, waiting times, customers turned away or otherwise cancelled, and server utilization. Experiments can then be made with different arrangements of servers to see which is the best.

Each simulation run needs to represent the behaviour of the queue over a period of time sufficiently long to cover any cyclic fluctuations in the arrival pattern. It must also be long enough to produce statistically stable results, as discussed in Chapter 8.

(b) Transport problems

The operations of a fleet of transport vehicles are often subject to the influence of unforeseen delays due to technical snags, bad weather, or other factors. If the fleet is providing a scheduled service, then a delay to one vehicle at one point in the transport network may cause many "reactionary" delays to services scheduled for that vehicle at other points. To reduce the disruptive effect of unforeseen contingencies, Management can provide standby vehicles, or insert time buffers in the schedule so that delayed vehicles have a chance of catching up with their schedule. The problem of determining how many standbys or how much buffer to insert, and where, is exceedingly complex, and is best studied with the aid of simulation. Frequency distributions representing the probabilities of various disturbing events can be included in the model, and the consequent disruptions to services computed under assumptions about fleet size, number of standby vehicles, and scheduling tightness.

A different problem arises in the situation where an "on-call" service is being supplied, as in the case of delivery to customers. In this case, in addition to unforeseen factors affecting the availability of vehicles, there is the unforeseen nature of the demand, which fluctuates from place to place and from time to time. The problem is to determine the best number of vehicles to have and the best rules for allocating vehicles to calls, so that they are met with least delays to customers and with a minimum of unproductive movements. This problem too can be fruitfully studied by simulating all the movements made in a typical period of time under specified assumptions.

(c) Mechanical handling problems Monday (-to see.)

Mechanical handling of goods at factories, warehouses, or docks often requires the use of several independent handling machines to perform different phases of handling. At docks, for instance, it is common for freight to be moved from warehouses to quayside by forklift truck or other mobile plant, thence lifted on board ship by crane. The problem that arises in such situations is that the movements of the various pieces of equipment are difficult to synchronise, with the result that often one piece is forced to wait while another completes its current task. This problem is especially acute if it is not possible for one machine to leave its load at some transfer store but must transfer its load directly to the next handling machine, as in the case of the container operations described in Chapter 11. Further examples occur in steel-works where a combination of

39

trucks and gantry cranes is often used to handle ingots from one operation to another.

The general problem can be formulated as that of determining the best (usually judged in economic terms) arrangement of handling machinery which is able to provide adequate capacity (volume of freight or workpieces handled per hour).

Due to the high degree of interaction between the various machines, it is usually very difficult to foresee the capacity of any given arrangements. It is, however, relatively easy to construct a model of the system and simulate the detailed movements of the machines, reproducing the interactions and interferences as they would occur in the real system. Then, by running the model over a long period of time, average working rates can be determined.

The main elements in this type of model are the machines. Events to be simulated are the journeys or activities that the machines perform, either singly or collectively (transferring a load from one machine to another would be a collective activity). It is usually unnecessary to consider explicitly the material being handled.

The duration of activities is normally variable. Often part of the variability can be attributed to trip length differences or other predictable factors. There is usually some residual unexplained variability which has to be treated by random sampling methods.

Due to the fact that there are relatively few machines in the typical handling operation, the number of separate entities needed in the model is usually quite small, compared with the requirements of the customer service model in which several hundred customer units may have to be simultaneously represented. Computer storage requirements are thus relatively undemanding. This partly explains why much of the original work in the field of industrial simulation was concerned with this type of problem.*

(d) Maintenance problems

Problems of queueing and storage also arise in connection with the maintenance and repair of equipment subject to random failure.

* K. D. Tocher's original work at United Steel in the late 1950's was performed on a Ferranti machine with a fast store of less than 100 locations.

The situation has much in common with the service of randomly arriving customers discussed in (a) above; failed machines can be regarded as customers arriving for service at the maintenance and repair facility. An important difference is that the population of customers is strictly finite, and the arrival rate at any given time depends on the number of machines in service at that time. Another difference is that part of the work load on the maintenance facility is likely to consist of routine preventive maintenance of equipment. Problems that arise are: how much capacity is needed for repair work? would it be better to invest in extra capacity, or in extra machines, bearing in mind that with less capacity there will be more queueing and hence a lower achievable utilization of equipment? Other problems may concern policies relating to preventive maintenance: how often should this be performed? should a life-expired machine be withdrawn from service when there are failed machines queueing for the use of common repair and maintenance facilities?

These problems do not admit of a simple solution, depending as they do on the interaction of many complex factors. A simulation model can be used to throw light on such problems by representing entries and withdrawals of machines from service over an extended period of time and showing how maintenance facilities, stocks of spares, and other factors are used, and how limited capacities affect the throughput of maintenance and repair work.

(e) Production buffer stocks

A frequently occurring problem in manufacturing is posed by the need to hold buffer stocks between two production units working at different rates. Chapter 12 deals with such a situation in a steel tube plant where the rate of tube-finishing work varied considerably from the rate of mill production of tube, necessitating a buffer store between the two processes (Fig. 6.1). Similar situations arise in mass

Fig. 6.1 Buffer stocks in a steel tube works

assembly work where the main assembly operations are performed on a continuous flow-line basis, but the subassemblies and basic part-making operations are worked independently of the line. Part and subassembly stores are then needed.

Storage problems also arise in the distribution of products to customers, after manufacture.

The general problem is to determine the most appropriate size for the store or stores, having regard to the nature of the input process and the output process, and the consequences of (1) filling the store and having to interrupt the input, and (2) emptying the store and having to interrupt the output. The rates of input and output usually fluctuate considerably within some overall production cycle (we are not considering jobbing production here) covering the full range of products made. It is thus far from straightforward to see what the peak storage requirements will be under different production arrangements.

The problem can be solved by using a model which simulates the movement of material into and out of the store or stores. When this is run over a sufficiently long period the pattern of stock fluctuations can be seen.

The model is very similar to those already discussed. The main events are the arrivals of material at the stores and the departures from the stores. The timing of these events is controlled by the schedule of work in the factory and the working rates of the various manufacturing processes. It is sometimes necessary to include the effect of work-breaks and manning arrangements in so far as these affect the rates of input and output from stores. Special events can be included to represent the commencement of these effects and subsequent termination, in a similar manner to the treatment of extraneous influences affecting customer service in the previous model.

It is often necessary to have quite complicated decision rules to represent the shop sequencing of work performed by foremen. Further rules are needed to determine actions when stores become either full or empty.

Model outputs consist of the time pattern of quantities stored, and delays or flow reductions caused by empty or full stores.

42

(f) Job shop scheduling

The term "job shop" or "jobbing production" denotes a type of production in which the work is of such a diverse nature that batching or flow lining is impractical. Rather the work must be treated as a series of individual jobs each with its own peculiar routing among the machines. At any one time in the factory there will be a vast number of jobs in various stages of progress. They will be in competition with one another for time on the available machines. There will thus be many possible ways of sequencing the work of the factory. The rules or principles used to decide on work sequence will affect the utilization of the plant as a whole, the volume of work in progress, and the punctuality with which work is completed relative to due dates quoted to customers.

The problem of determining the best rules to use is clearly a formidable one, due to the complexity of the operations and the number of criteria which have to be satisfied. One approach which can be used is to simulate the working of the job shop under different rules, and inspect the results. Several attempts have been made to do this, but to date only simplified cases have been studied in this way. Actual operations have mostly proved too difficult to model in full detail.

Models

We now consider how to construct models in order to solve the above problems on the computer. We first observe that these problems are all concerned with systems of discrete entities, such as vehicles, moving from place to place. For instance, in the customer service problem we are concerned with the arrivals and departures of customers, the way they join queues, and the movement of individuals within queues to reach service points. In transport and handling problems we are concerned with movements of vehicles and handling machines and the passengers or cargo they convey. In maintenance problems we are concerned with the entry and the withdrawal of machines from service, and with the movement of machines through various stages of maintenance and repair activity. In buffer storage problems we are concerned with movement of stocks into and out of

43

store. Lastly, in the job shop problem we are concerned with the movement of jobs through the machine shop.

One approach to this type of problem is to define all the possible states that the system can attain, and write down mathematical expressions for the probability of each state in terms of the probability of adjacent states and the probability of transitions from these. A solution to these equations, if achievable, will give the values of all the state probabilities and hence a complete solution to the problem.

However, in most of the situations described above the possible number of states is enormous. Moreover there are an enormous number of ways of reaching any given state from adjacent states. Hence this method becomes impossibly cumbersome.

The alternative is to play out all the movements in time sequence and record the occurrences of the actual states of the model system as the run progresses. This is of course the simulation approach. The approach can be regarded as an extension or generalization of the approach adopted in the last chapter for following the tracks of subatomic particles through the maze of atoms inside a nuclear reactor. The main difference is that it is no longer possible to simulate entity (particle) paths independently because interactions, such as those between members of the same queue, are usually important. We cannot therefore take the entities one at a time and simulate their entire histories; we have to move all the model entities together so that interactions can be correctly represented. This poses a problem because of the sequential nature of computer calculations. A clock has to be set up recording model time, and all necessary movements simulated at each clock time in the run. By advancing the clock stepwise through the period of simulation all movements can then be simulated and all interactions correctly preserved. A feature of this type of model is that times of entity movements can be calculated in advance, either by table look-up in the case of an entity undergoing a deterministic activity, or by random sampling in the case of an entity making an unforeseen movement such as customer arrival at service point. The clock can therefore be advanced from event moment to next event moment without halting at intervening times when no activity can occur, since future events can be pre-timed. By an event is meant any change to any model entity.

Alternatively a fixed time-step can be employed, but this method is generally wasteful of computer time, because additional time-steps are introduced which are unnecessary.

The scheme for the event-to-event method of time-advance is illustrated in Fig. 6.2. There are several possible ways of determining

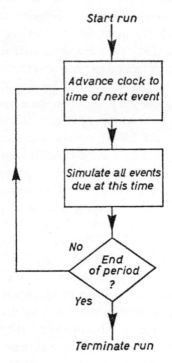

Fig. 6.2   General scheme for event-to-event model

when the next event will be due and what other events will also be due at this time. The most straightforward method is to have a list of current and future events stored in the computer, together with due-times. This can then be scanned when a time-advance is required to determine the next event time. Another possibility is to attach a next-change time to all model entities, and advance the clock to the earliest of these times. Whichever scheme is adopted, the model-builder has to be careful to see that spurious effects are not introduced by the order in which simultaneous events are simulated. Another problem is to ensure that new events called in train by a given event are not overlooked in the event scan, and the clock advanced without their being executed.

These problems need not concern the model-builder if he uses a computer programming language designed specially for simulation work, because an event-sequencing routine will be provided. It is then only necessary to specify what events are to be included in the model, and the event-sequencing routine of the programming language will ensure that these are simulated in the correct order when the model is being run on the computer. A list of the commonly available simulation languages will be found in the Appendix.

Let us now consider how an event-to-event model would be constructed for a typical customer service simulation model. The main event types would be (1) customer arrival, (2) service commencement, and (3) service completion. In some models it might be necessary to include other events to represent the influence of outside factors on the availability of customers or servers. For instance, when studying port operations it might be necessary to consider changes in the state of the tide because these influence the arrival and departure of ships. The state of the weather may also have to be taken into account. Such effects can be incorporated by including special events to represent the onset and termination of the influences, and simulating them at appropriate times in the run.

Customer arrivals would be generated by sampling from appropriate inter-arrival interval distributions. Where in the real-life situation arrivals are scheduled, times could be built in and actual times simulated by applying a random disturbance to represent occasional departures from plan.

Service times would likewise be determined by random sampling. It might be necessary to use different distributions for different classes of customer and server.

The model would contain rules representing how customer and other demands on servers' time would be dealt with in the wide variety of situations that could arise. These rules would be incorporated in the blocks of program specifying the changes of state caused by the taking place of each type of event. Thus event type (1), customer arrival, would contain rules governing to which free server the customer would be allocated, or, if no server was free, to which queue.

Models of transport situations, maintenance, storage, and job shop operations are constructed along very similar lines.

APPLICATIONS

(A) Transportation

(1) *Air transport*

Many airlines, including BEA, Air France and Air Canada, have studied factors affecting the punctuality of their aircraft by means of simulation. The main inputs to their models are:

(1) technical snags to aircraft;
(2) incidence of bad weather periods;
(3) maintenance checks.

The main outputs are punctuality at different points on the network, utilization of aircraft, and the use of standby aircraft.

United Air Lines has constructed a model of a Maintenance Station, representing the arrival of aircraft for checks and the carrying out of maintenance activities on each check under given assumptions concerning hangar space and available manpower. The outputs of the model are trip delays caused by aircraft returning late from maintenance station, and demands on reserve aircraft. Some airlines have studied passenger handling at airports to see how many check-in desks and other facilities are needed to provide adequate service, measured in terms of queueing times.

Many studies of air traffic congestion have been undertaken by Air Traffic Control Authorities. Both ground delays on taxiways and at runways holding points and air delays in landing stacks have been studied in relation to planning decisions for future runways and other facilities.

Maintenance and repair of equipment have been studied by several airlines. Overhaul of major spares in workshops has been studied by BEA to determine benefit of alternative priority rules.

United Airlines, BOAC and BEA have studied the repair of aircraft engines to see how many spares should be held at outstations, and to determine the load on central maintenance and repair facilities. Inputs are engine removal rates by station, shipping delays between stations, and days required for overhaul. Outputs are number of engine shortages by station and duration of these, and enforced idleness of men in shops due to lack of work.

(2) *Shipping*

Many studies of port operations have been carried out using simulation to determine the number of berths needed. A leading oil company has studied one of the Middle East crude ports. The Port of London Authority has studied the required capacity of locks at the entrance to some of the London docks. A study of container handling in ports to determine the best handling arrangements and machinery has been carried out by National Ports Council, on behalf of British ports in general.

The U.S. Dept. of Commerce has studied shipping on the Great Lakes with simulation to determine the optimum number and location of pilots needed.

The operations of several U.S. west coast ports have been studied by the Department of Commerce to see how well an additional military load could be handled.

(3) *Railways*

Movement of trains on part of the Boston–Washington route has been studied by the Department of Commerce with the object of determining the utilization of track, and periods when maintenance would be possible.

(4) *Road*

Many studies have been made of the movement of road vehicles in the centres of cities. Some of these have been aimed at finding out how to improve existing street networks; others have sought optimum control rules for use with computer-controlled traffic schemes. In both cases simulation models have been used to compare various control systems such as fixed-time, vehicle-activated, and minimum-delay in terms of predicted behaviour either at a single intersection or on a network of streets under different flow conditions. Simulation has also been used to study operations at busy bus terminals.

(B) Manufacturing

Several models have been constructed to study storage problems in manufacturing plants with the object of determining the effects of changed equipment and methods on storage requirements for part-finished material. Many studies of handling requirements at steel-works and other plants have also used simulation.

(C) Communications

Studies have been made of communications networks to determine the extent of queueing of calls, and calls lost during peak periods. Situations studied have ranged from underground communications in a colliery of the National Coal Board to real-time computer systems with remote entry of inquiries, or of jobs for processing.

(D) Mining

The National Coal Board has studied different methods of repairing coal-face machinery. The problem was regarded as a queueing problem where the "customers" are machines that have recently broken down, and the servers correspond to overhaul facilities at remote workshops. The object was to find the number of machines needed to give minimum costs.

(E) Public Utilities

Storage of water for irrigation, power production, flood control, and recreation has been studied by several Water Resource Authorities in the U.S.A.

References

[1] K. D. TOCHER, *The Art of Simulation*, English Universities Press, 1963.

[2] T. H. NAYLOR *et al.*, *Computer Simulation Techniques*, Wiley, 1966.

[3] D. CHORAFAS, *Systems and Simulation*, Academic Press, 1965.

[4] H. B. MAYNARD ed., *Industrial Engineering Handbook*, 2nd edn, McGraw-Hill, 1963.

Chapter 7

Models of combat

The playing of war games has formed part of the training of military commanders for many years. More recently these games have been used to provide a means of assessing weapon effectiveness. Such simulations provide the best means of assessing the overall effectiveness of future weapons or weapon systems, in the context in which they will actually be used. In particular they enable the influence of terrain to be included, a factor which plays an important part in any ground or low air battle. By using natural terrain features, forces are able to conceal themselves from the enemy's fire. The effect of such concealment on the outcome of the battle can only be properly assessed by reconstructing typical battle sequences in a war game. These games can be played manually or on a computer. The manual method allows more realistic rules to be used because experienced serving officers can be asked to play the part of the two side's commanders. They bring their experience of battle conditions to the game and handle the weapons accordingly.

The disadvantages of the manual method are, firstly, that the game's progress is very slow, usually much slower than real time; this means that the commanders have more time than they should for making decisions, which are somewhat unrealistic in consequence. Secondly, very few games can be played, due to shortage of time. Thirdly, the games are not repeatable. For these reasons, the trend is to put the war game on to a computer. In such automated games, or computer simulations as we shall henceforth call them, the tactics that the commanders will employ are codified at the outset and built into the model as a set of rules.

The field of combat may be air, sea, or land, and the scale may vary from a few metres to several hundred kilometres over which the battle is being fought. The criterion of effectiveness of a force may be the number of enemy units destroyed or immobilized, or it may be territory gained. Another criterion may be the degree to which the enemy's advance is delayed by the action of a defending force. Such delay may enable the defender to summon up reserves.

The feature that most distinguishes the combat model is the fact that there are two opposing mutually destructive forces. It is customary to denote these by Red and Blue, as illustrated in Fig. 7.1. The model is shown as a computing box taking inputs of the equipment and weapon characteristics of each side, and calculating the outcome, taking into account prevailing environmental conditions of terrain and weather which are assumed common to the two sides.

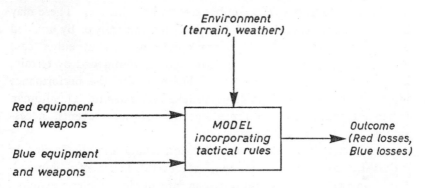

Fig. 7.1   General scheme for combat model

The tactical rules used to decide on the movement of units and the selection of targets to attack are shown incorporated in the model. In theory these could be made variable and regarded as an extra input, but in most practical computer simulations they are taken as fixed, and it is the weapons that are the main object of study.

The quality of the commander's information, the quality of his decision-taking, and the effectiveness of communication on the field are important, often crucial factors. The flow of messages can be simulated in the model, with random interruption due to "noise"; sightings of the enemy can also be treated in probabilistic terms.

E

Sometimes it is necessary to carry out independent simulations of message flows in order to isolate potential bottlenecks in the communications system.

Combat units, event structure and time advance

The basic entity is the combat unit. This may be a tank, an aircraft, or a ship, according to the type of battle. The combat unit can move, usually in two dimensions (in the case of submarines and some aircraft studies three dimensions are needed), on the battlefield. It can sight enemy combat units, or be given intelligence of the whereabouts of enemy units. It can fire weapons at enemy targets, which may be other combat units like itself, sighting devices such as radars, or communication lines.

The basic events of any combat are firstly sightings. These may be made by means of detecting devices such as radar, or by unaided eye or ear in the field. The main determinants in either case are (1) range, (2) whether the line of sight is unimpeded by terrain, and (3) meteorological conditions. There is also the performance of the radar operator, or lookout, to take into account. This human factor introduces a chance element into sighting events, necessitating the use of random sampling.

The second type of event is decision-taking and the issuing of orders by commanders, perhaps at several levels of command. Based on new sightings and previous information, the commanders decide on movement of units and the releasing of fire, the two other main event types.

The third type of event is physical movement of a combat unit. This may be called for by the unit commander, in pursuit of or in retreat from an enemy unit, or it may be called for by the force commander as part of marshalling or for some other tactical reason. A movement once decided on can usually be treated as deterministic.

The fourth type of event is the releasing or firing of offensive weapons against enemy targets. The outcome of this event is subject to several unpredictable factors requiring random sampling. Firstly, the accuracy of aim is variable, and the weapon may fall on or away from target accordingly. Secondly, the actual destructiveness of the weapon may be variable, even when on target, according to angle and velocity of impact and so on. A number of other factors may be considered important and hence included, such as whether the unit concerned is moving or stationary, whether visibility is good or bad, etc.

These probabilities are of course difficult to measure. Field trial data can be used to give an indication, but even a figure so derived cannot be regarded as highly accurate, because actual battle conditions are so unlike the trials, the quality of the enemy's protective armour can only be guessed at, and the number of actual firings (especially if the weapon is expensive) is usually inadequate for a reliable statistical estimate. For new weapons there may be no data, and judgement must be relied on.

Such arguments do not of course render simulation exercises valueless, because even a guessed figure for kill probability is better than none. Moreover, sensitivity analyses can be made with the model so that the sensitivity of battle outcomes to variations in the figure used for kill probability can be measured.

Since, in general, both sides can make moves in a manner unpredictable to the other side, it is not generally possible to foresee when the next event will take place. For instance, we cannot predict when new sightings will be made, or other events which will set new events in train. Thus we cannot in general assign a next-event time to each unit as we can in the models discussed in Chapter 6, but must resort to a time-stepping method of advancing the model clock.

Some models are concerned with evaluating the effectiveness of defence against a given attack pattern. In this case the movements of the enemy (attackers) are known in advance. It may then be possible to foresee the next event for each defending unit, in which case a critical-event method of time advance can be employed.

Representation of terrain

One of the most important factors in land combat is the nature of the terrain over which it is fought. Unfortunately terrain is one of the most difficult features to include realistically in a simulation model. What is needed is a method of determining within the computer which combat units, and which observers, are in line-of-sight with one another. Ideally we would like to be able to specify the general nature of the terrain—whether it is flat or hilly, how wooded it is—and arrange for the computer to generate a typical portion of terrain (in terms of intervisibilities between pairs of points) which could be used in the simulation. But this is a formidable task due to the correlations that exist between heights of neighbouring points.

A way out of the difficulty is to take an actual area of ground, representative of the type of terrain which is of interest, and use survey data on ground heights and woods as basic input data for the model. If the combat area is divided up into a number of squares and each is assigned the mean height (trees included), then by simple three-dimensional geometrical considerations the squares visible from any given square can be calculated.

This method has the disadvantage that, for accuracy, the squares must be made small so that there is not too much height-variation within each. This will require a large table of height data for storage in the computer, which necessitates either the use of a computer with a large immediate access store, which is expensive, or holding the table in slow-speed backing storage such as magnetic tape or disc, which slows the running of the simulation.

There is one type of combat situation where the terrain *can* be synthesized and the above disadvantages avoided. This is where the attacking combat units, such as low-level bombers, can be treated independently of one another, and the study is concerned with how many attacking units can penetrate the defences without being observed from a given vantage point.

In this case we can divide the area surrounding the vantage point into a series of circular annuli. In each, we form a probability distribution of "critical altitudes", i.e. the altitudes below which the attacker is hidden from the vantage point. This can be done by taking map data in the type of terrain which is of interest, and measuring critical altitudes at a sample of distances from a sample of points on the map.

We can then generate the critical altitudes along the path of the attacking unit by successive random sampling from the annuli distributions, in such a way that correlations are preserved. Quite realistic samples of critical heights can be generated in this way,

54

which compare well with samples of actual data, in the statistical sense of giving the appearance of coming from the same parent distribution. Descriptions of this type of terrain generator are to be found in the literature [1].

Another approach to the problem of representing terrain is to deal with the major terrain features only, on the assumption that it is these that effectively determine intervisibility, cover, and ease of movement. These features, e.g. ridges of high ground, areas of wooded country, and marshes, can be regarded as effectively independently distributed over the terrain. Thus, to generate a typical portion of terrain it is sufficient to sample at random from distributions based on the actual frequency with which these features occur in the real-life terrain.

Apart from its effect on intervisibility, terrain is also important for the movement of land units. Thus it is necessary to include a factor to represent the "going" at each point of the simulated terrain. Also the steepness of hillsides affects the maximum speed of land units.

The actual approach adopted to the problem of representing terrain will depend on the particular type of problem being studied. In some problems, such as protection against attack by high-flying aircraft, terrain is not relevant.

APPLICATIONS

Most of the work in this field is classified, and only an outline of each application can be given.

Land combat

Several models have been constructed of small-scale land combats between small forces of armed vehicles [2]. The object of these is to study the effectiveness of the weapons used in a given terrain and under given enemy action.

In some models the terrain is represented by a grid of squares, each having a mean height above sea level, and other characteristics, such as whether wooded or not. Before simulation, a matrix of intervisibility is constructed, giving the squares that can be seen from each given square. This matrix is then referred to each time a question of intervisibility occurs in the simulation.

Time is advanced in fixed steps, and at each time step two events are possible for each combat unit. Firstly, the unit may decide to move to an adjacent square (eight possible choices) in order to move towards some terrain objective or to avoid enemy fire. Secondly, the unit may decide to fire on an enemy unit which may be damaged or destroyed.

Communication between units is included. Messages take time to transmit and sometimes have to be repeated. The outcome of the combat is measured in terms of number of enemy units destroyed in relation to the number of friendly units lost. The effectiveness of a given type of unit is measured in terms of the number of enemy units destroyed per friendly unit, averaged over a series of runs of the simulation. Thus different types of unit and weapon characteristics can be tried out in the model, and compared with one another.

One of the difficulties encountered in this type of model is in constructing the rules of movement, since this involves considering what terrain objectives each side would have, and how they might be modified during the battle by the course of events. In one model a probabilistic approach was adopted at first, wherein squares adjacent to the square occupied by an enemy unit were assigned weights representing desirability. This approach was later abandoned because it produced unrealistic movement patterns. Another problem is posed by the need to look ahead when planning movement. Otherwise units, while moving towards objectives, could find themselves trapped in culs-de-sac.

Air combat

Most models in this field are concerned with systems of defence against attack by enemy aircraft, rather than combat between groups of aircraft.

The case of attack by low-flying aircraft has been extensively studied, as this is currently thought to be a method likely to be used by potential enemies. By flying low, aircraft can use the terrain to conceal themselves from defensive radar observation, and with the development of radar chains this is about the only form of concealment now available to an attacker. A flexible model for studying the effectiveness of defence systems against this form of attack has been developed at the SHAPE Technical Centre [1].

Combat units consist of the attacking and defending aircraft, air-to-surface missiles, and anti-aircraft gun batteries. Principal events are (1) detections of attacking aircraft, (2) launching of defensive aircraft, and (3) initiation of other defensive weapons. All events are tied to combat units. Any given combat unit will normally have a next event scheduled.

Control of the program alternates between a time-advance phase, which consists of a step to the next event on the event list, and a phase of event execution.

Movements of the attackers can be treated as fixed events scheduled in advance. Such pre-scheduled movements are placed on a separate list which is read in at the beginning of the run, and scanned during the time-advance phase.

Due to varying detection probabilities with time, detection events cannot normally be foreseen in the model (time to detection does not conform to any simple probability distribution), so a fixed time-step must be used to determine detections. This can be arranged via the pre-scheduled events list.

Treatment of terrain

Terrain information is taken from surveys of actual sites of the defensive weapons, so as to make the representation as realistic as possible. A terrain "mask" is derived from this information, showing the maximum possible aircraft detection range in any direction.

Combat at sea

Less work has been done in this area. Some studies have been made of the effectiveness of different submarine manoeuvres to avoid or frustrate attack by surface ships.

References

[1] S. H. HOLLINGDALE, *Digital Simulation in Operational Research*, English Universities Press, 1967.

[2] R. E. ZIMMERMAN, "Simulation of Tactical War Games", in *Operational Research and Systems Engineering*, ed. C. D. Flagle, Johns Hopkins University Press, 1960.

Chapter 8

Design of simulation experiments

The problem of estimation

When we use simulation to solve a problem, the answers we get are not unique if any sampling has been involved, because the calculation then depends to some extent on the actual pseudo-random numbers chosen. There is nothing special about these numbers; another set could equally well have been used, but this would have given slightly different results. We thus have the paradoxical situation that for any set of conditions to be simulated, there is not one single result but a range of different, equally valid results possible.

This situation arises from the fact that we are representing factors that are to some extent unpredictable. We must therefore expect some uncertainty in the results. It is clear that the uncertainty diminishes as the period of time over which the results are averaged becomes longer; therefore, if we simulate long enough, we can reach a situation where an answer is obtained which, even though not perfectly precise, is good enough for practical purposes.

This situation is akin to that found in statistical sampling work, where the object is to measure some attribute or characteristic possessed by members of a large population by taking measurements on a sample only. Such a procedure yields only an approximate measure or "estimate" of the attribute or characteristic in question. As the sample size is increased, the precision of the estimate is improved, but perfect knowledge is only attained when the sample contains the whole population. Since the object of sampling is to avoid dealing with the whole population—which would often be infeasible and is nearly always uneconomic—we must accept some degree of uncertainty in the estimates obtained. We need to attach some measure to this uncertainty so that we can design samples which give the precision required in any given case.

In carrying out simulation studies we likewise need to attach some measure to the uncertainty of the results. We can do this, for a given run, if we break the period simulated into several subperiods, and regard the results obtained in each as coming from a population of results obtainable from similar subperiods. The general problem is then to determine the precision of estimates obtained from a run made up of several successive periods of simulation. This situation resembles that normally found in sampling work, except that successive periods possess some degree of mutual correlation, due to the fact that the starting conditions of each period are the finishing conditions of the previous period. This correlation can often be virtually eliminated by making use of natural periods in the system being simulated. For instance, in an air traffic model such as that described in Chapter 10 use can be made of the natural lull that occurs in the real-life system in the small hours of the morning. If 24-hour periods commencing at dawn are taken, then starting conditions will be very little affected by activities in previous periods, and the individual periods can be assumed to be independent of one another.

In models containing no natural periods, some degree of correlation will undoubtedly exist. It is possible to allow for the effect of this by using a correction factor based on the degree of correlation. This subject is, however, outside the scope of the present volume. In what follows we will assume it is reasonable to consider that the simulation run can be broken up into a set of periods, the results from which can be taken as independent values sampled from some unknown parent population.

Distribution of sample estimates

Let us assume, then, that we are going to estimate some characteristics of the population by taking a sample of n values (or an "n-sample"), and calculating the mean. To see how precise such an estimate is, we study the distribution of means of n-samples in general. We denote the mean and variance of the parent population by μ and σ^2, respectively. Further, we denote the first value drawn in a sample by x_1, the second by x_2, and so on up to x_n. The sample mean can then be written:

$$\bar{x} = \frac{1}{n}(x_1 + x_2 + \ldots + x_n)$$
$$= \sum_i x_i/n$$

Hence the sample mean, regarded as a variate, can be expressed as the sum of the n variates x_i/n, $i = 1, 2, \ldots, n$. These variates each have mean μ/n and variance σ^2/n^2.

Now the mean of the sum of several independent variables is equal to the sum of their means, and likewise the variance of the sum is equal to the sum of the variances. Thus

$\bar{x}$ has mean $n(\mu/n)$, i.e. μ, and variance $n(\sigma^2/n^2)$, i.e. σ^2/n.

Confidence limits

The standard deviation of $\bar{x}$ is thus $\sigma/\sqrt{n}$. This standard deviation is a convenient measure of sampling error and is called the *standard error*.

It can be shown that the distribution of $\bar{x}$ approximates to the Normal distribution whatever the distribution from which the sample values are drawn, provided only that n is sufficiently large.

Let us now consider the range of values that are likely to be obtained by drawing random values from this distribution—see Fig. 8.1. We can say that on most occasions such values will lie

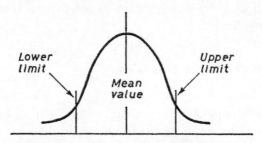

Fig. 8.1 Distribution of sample estimates

somewhere near the centre of the distribution. If we set two limits equidistant from the centre, such that 95 per cent of the area is enclosed by them, then we can say that on 95 per cent of occasions the value drawn will lie between them, and using the known shape of the Normal curve, we find that these "95 per cent limits" are at a distance approximately twice the standard deviation on either side of the centre of the distribution. These limits measure the likely error inherent in estimates made with samples of size n. We can use them to assign plus and minus limits to such estimates.

The procedure for determining error, or confidence, limits is thus as follows:

(i) From the sample values $x_1, x_2, \ldots, x_n$ estimate σ, the standard deviation of the population.

(ii) Divide by $\sqrt{n}$ to obtain an estimate of the standard error of sampling.

(iii) The required confidence limits are now the points at a distance twice the above standard error on either side of the mean value of the sample.

Relationship between precision and sample size

The precision of an estimate depends directly on $\sigma/\sqrt{n}$, where σ is the standard deviation of the population and n the size of the sample. It follows that precision is inversely related to the square root of the sample size. This implies a law of diminishing returns in

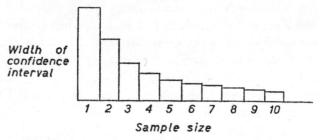

Fig. 8.2   Relation of precision to sample size

relation to sample size, as illustrated in Fig. 8.2. As the cost of computing is more or less directly related to sample size, there comes a point where the cost of further running is not justified by the increase in precision obtained.

It can sometimes happen that simulation results are only meaningful when applied to a relatively short period of time, due to the fact that the conditions assumed in the model only persist in the real-life situation for a short period. If so, this period may be shorter than that needed to give reasonable precision in the results.

What this situation indicates is that our degree of knowledge concerning the working of the system and all its interactions is not sufficient to enable us to predict its behaviour with sufficient precision over the given time-period. We must either replace some of the randomness in the model with deterministic mechanisms (in other words, explain some of the unpredictable effects and insert their underlying causes in the model), or be content with obtaining a range of probable results rather than a single result.

Variance reduction techniques

So far we have assumed that the sample values are independent of one another. It is possible, however, to improve the precision of a given sample by forcing a degree of dependence on values by careful choice of pseudo-random numbers used. This can be achieved by the technique of *antithetic variables*, which consists of performing two runs, identical apart from the choice of pseudo-random numbers, and averaging the results. The two sets of pseudo-random numbers are chosen in such a way that they have a strong negative coefficient of correlation. This means that the two runs will tend to produce results on opposite sides of the population mean, and, when taken together, will give a result closer to the mean than would be likely otherwise.

The first set of random numbers is produced in the usual way with a standard generator. Suppose this set is $x_1, x_2, \ldots, x_n$, and that, for the sake of argument, all the x's lie in value between 0 and 1. Then the second set is derived from the first by taking complements with respect to 1 as follows:

$$y_1 = 1 - x_1; \qquad y_2 = 1 - x_2; \ldots \qquad y_n = 1 - x_n,$$

to give the set $y_1, y_2, \ldots, y_n$. The effect of this is to force values to be drawn from opposite ends of distributions in the two runs, so that the results tend to be negatively correlated, thus achieving the desired improvement in estimating precision, mentioned above.

Control variates

In some situations it is possible to compare the model with a simplified but similar model which can be solved mathematically. For instance in a queueing problem, it may be possible to compare the model with one of the standard models whose analytic solution is known.

It is then possible to estimate the size of the error due to sampling by simulating the simplified model, using identical sampling numbers, and then comparing the result with the known theoretical result. This technique is known as using a *control variate* because the simple simulation can be regarded as providing a statistical control on the complex simulation (the output from this simple simulation plays the part of a control variate).

Comparisons

When a simulation model is being used to make comparisons among several alternative real-life systems, or different levels of the controllable variables, then a technique similar to the control variate technique can be employed to reduce the variability of the comparisons. The same set of pseudo-random numbers can be used for each of the runs.

The actual values chosen from the sampling distribution will then be identical in each of the case studies. Genuine differences due to the effect of having different controllable factors will be more obvious because they will not then be obscured by spurious sampling differences.

In this connection it is important to use a separate stream of random numbers for each distribution, so that an identical sequence of values is sampled from each distribution in every run.

It is not necessary to have separate generators, as separate streams can be produced from the same generator merely by using different starting values. Care must be taken, however, to ensure that these starting values (sometimes called seeds) come from widely separated parts of the random number sequence, otherwise unwanted correlations may be produced in the model which may lead to quite erroneous results.

If recommended starting values are not supplied with the generator (and usually they are not) then the user should generate the whole, or a large part, of the pseudo-random number sequence, and choose a set of starting values by taking, say, every 10,000th number in the sequence.

References

[1] C. E. WEATHERBURN, *A First Course in Mathematical Statistics*, Cambridge University Press, 1949, Chapter 6.

[2] G. U. YULE and M. G. KENDALL, *An Introduction to the Theory of Statistics*, Griffin, 1965, Chapters 16, 17, 18.

PART 2

CASE STUDIES

Chapter 9

Provision of standby aircraft

This study was undertaken for the scheduling and fleet-planning branch of an airline operating a fleet of about one hundred aircraft. Traditionally this airline had held in reserve a number of its aircraft for use in case of aircraft shortage caused by technical trouble or bad weather. Due to the rising capital cost of modern jet aircraft, it became important to confirm that these "standby" aircraft were really needed. It seemed likely that a simulation study would be helpful, and this led the scheduling and fleet-planning branch to call in the Operational Research Section to advise them on the suitability of such a study.

After about two months of study the problem was seen as essentially a queueing problem. The customers were identified as the flights on the airline's published timetable, and the servers were identified as the aircraft in the operating fleet, including standby aircraft. Due to technical snags or bad weather, the number of aircraft available for service (including standby) could on occasion be reduced below that required to operate the schedule, and then flights would be delayed awaiting aircraft.

It was felt, however, that analytical formulation in terms of queueing theory would not give an accurate solution due to the complex nature of the flight schedule. It was therefore proposed to construct a simulation model having as basic input the number of aircraft in the fleet, together with the schedule of flights required to be flown, and having as basic output the delays experienced by these flights. The effect of increasing or reducing the number of standby aircraft could then be determined by performing runs of the model with different numbers of aircraft in the fleet, and seeing how the punctuality varied.

67

F

We now describe some of the simplifying assumptions that were made. These were by no means trivial in their effects on the result, but were well worth while in enabling approximate answers to be obtained in the time available.

Allocation of aircraft to flights

The allocation of aircraft to flights is in practice a complex manual scheduling operation performed by several experienced controllers. In making their allocations, the controllers look ahead several days to ensure that aircraft are correctly positioned for future flights and maintenance checks. They also take into account for each flight the hours involved, and for each aircraft the hours remaining before the next maintenance check, in order to ensure that aircraft neither exceed the time allowed before check, nor arrive with hours unflown.

For purposes of the simulation study it was felt to be unnecessary (even had it been possible) to reproduce the allocation procedure in every detail, because most of the factors considered by the controllers could be ignored on the basis that they would seldom if ever cause flights to be delayed. The requirements of punctuality would nearly always be put before other considerations. It was possible therefore to put a simple first-in, first-out (FIFO) rule into the model to cover the flight allocation aspect.

In retrospect this simplification seems obvious, but at the time, it took several weeks to make and justify.

Weather effects

It was necessary to study the effect of bad weather as this was held (by the airline management) to be a major cause of aircraft shortage. Records showed that while the movement of aircraft en route between airports was seldom seriously affected by weather, take-offs and landings at airports were often delayed by periods of bad visibility. It was decided therefore to include a representation of such periods in the model, together with a set of rules to represent the reaction of the airline to these conditions when they occurred.

At first sight, it seemed easy enough to use a random sampling method to generate bad visibility periods in the model. However, on further consideration such an approach was ruled out because of the difficulty of reproducing correctly correlated patterns over the whole network. Bad weather at Jersey, for instance, is nearly always associated in real life with bad weather at neighbouring Guernsey. Another important correlation is with time of day. Bad visibility is much more likely in the early morning than at any other time of day. To ignore these correlations which tend to bunch periods of bad weather would give an inaccurate account of the overall effect of bad weather on the flight schedule. A fleet size that gave adequate protection against uncorrelated periods might prove inadequate to cover aircraft shortages produced by correlated periods.

The problem was solved by using actual weather patterns as recorded in previous years, covering several months of actual weather.

Technical snags to aircraft

A study of the incidence of technical snags was made. It was realized that while faults could occur at any time, their influence was only felt when they were reported, and this was most likely to be done at the following times: on landing or prior to take-off, by the aircraft's captain, or during pre-flight checks by ground engineers. This meant that the possibility of a snag "occurring" to delay an aircraft could be restricted in the model to (1) immediately prior to aircraft departure, and (2) on aircraft arrival. At these moments random sampling could be used to determine whether a snag had occurred and if so how long repairs would take.

Accordingly, distributions of repair durations were collected from engineers' records, for each aircraft type in service. It had to be assumed that these distributions would validly represent repair durations in the future, possibly under conditions of greater urgency. This assumption was arguable. There was some evidence to suggest that when aircraft were in short supply snags were "carried" by aircraft captains in order to maintain punctual services. But it was impossible to ascertain the extent, if any, of this effect from looking at recorded data. A controlled experiment, which was outside the bounds of the study, would have been needed, and the results of such an experiment would have needed very careful interpretation.

Other causes of aircraft shortage

Causes of aircraft shortage other than bad weather and technical snags were identified as follows:

(1) aircraft late from maintenance checks;
(2) lengthy flight duration due to head winds en route;
(3) delays to flight departure due to delays in loading passengers, baggage, or freight into the aircraft.

These were all included in the model, and represented by random variables. Maintenance checks were regarded as specially long flights, with durations sampled from a special distribution of check durations. Maintenance activities were not represented explicitly.

The model

The fact that it had been assumed that flights could be allocated to aircraft on a strict FIFO basis meant that the following very simple scheme could be used for the simulation model.

The flights on the schedule were held on a magnetic tape and each one read into a "current list" of flights when the model clock reached scheduled departure time. At each time-advance the model scanned the current list in order of arrival and looked for suitable aircraft for each flight on it. On finding one, the complete flight was simulated. In other words, events consisted solely of flights arriving (being inserted into the current list) and flights being carried out. In queueing terms these events can be interpreted as (1) arrival of customers (flights becoming due) and (2) service of customers (aircraft performing flights). This illustrates the queueing nature of the model.

For simplicity, the fixed time-step method was adopted, the chosen interval being five minutes, which was the smallest unit of time recognized in the flight schedule.

Figure 9.1 shows the treatment of the flight event. First the state of the weather was inspected at the origin. If clear, the next step was to find an aircraft. Then the possibility of a snag was considered, by random sampling. If one was found, the duration of repairs was determined by further sampling. If the duration found was less than 30 minutes, processing of the flight was continued,

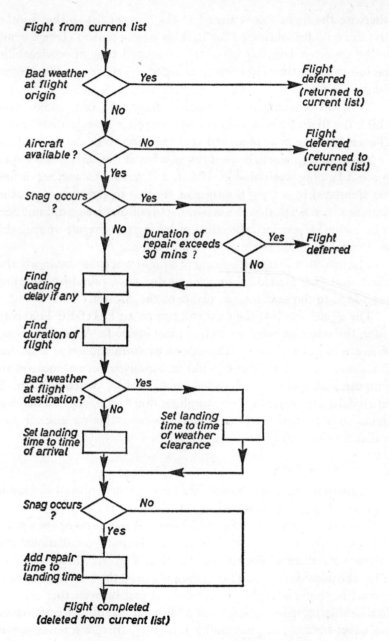

Fig. 9.1   Logic of a flight event

71

otherwise the flight was returned to the current list so that another aircraft could be considered for it. This was because a change would not be considered in real life if the estimated time of serviceability was less than the time (30 minutes) needed to prepare another aircraft to take the flight.

Then the possibility of a loading delay was considered, after which the flight time was found by sampling from a distribution. This enabled the time of arrival at destination to be determined.

The next step was to inspect the weather at the destination airport to see if landing was possible. If not, a diversion to another airfield was simulated by setting the time of landing to the time of weather clearance. It was felt that the actual details of diversions did not need to be included, provided the effect of having the aircraft unavailable for a suitable period of time was allowed for.

The final step in the processing of a flight was to ask again whether a technical snag existed. If so a repair duration, found by sampling, was added to the next time available of the aircraft.

The model was programmed and run on an EMIDEC 1100 computer, the program being written in machine code. A fixed time-step of five minutes was used. The speed of simulation was such that about one week of real time could be simulated in an hour on the computer, although speed varied considerably, depending on the actual delays generated in the sampling, due to the fact that deferred flights were reconsidered at every time-step, whether aircraft were available to take them or not.

Use of the model

Validation runs, simulating past performance, proved the model to be an adequate representation.

The first use to which the model was put was to compute a graph of flight delay-time against fleet size by running simulations with different numbers of aircraft in the fleet. Fig. 9.2 shows the form of the relationship found. This curve measured the benefit of standby aircraft in terms of flight punctuality. It can be seen that a law of diminishing returns applies, each additional standby aircraft having less effect because, as punctuality improves, there are fewer opportunities for further improvements.

This series of points was computed principally as a demonstration of what could be done with the model. The Operational Research Section were then asked to provide an answer to a practical question which had just arisen in connection with planning the deployment of the fleet for the forthcoming summer season. This was to determine the best location for one of the existing standby aircraft. Three possible airports were being considered, none of which seemed to have a clearcut claim to the aircraft, although all wished to have it. Of the three, which we will call A, B and C, A and B had a very similar frequency of arrivals and departures, while C with a much higher movement frequency already had one standby aircraft allocated to it. Accordingly the model was run to simulate the operation of the flights with the additional standby aircraft located at A, B and C in turn.

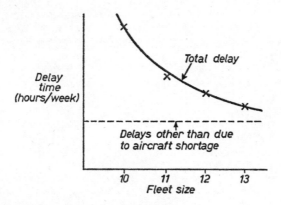

Fig. 9.2 Influence of airline fleet size on delays

The interesting result of this series of runs was to show that there was little to choose between the alternatives. Evidently aircraft shortage affected A, B and C about equally, because punctuality improved by much the same amount of each of these places when the extra standby aircraft was placed there. It was somewhat difficult to interpret the results meaningfully because the number of serious delays in each case—the main criterion—was small and liable to sampling fluctuations.

Subsequent uses and development of the model

After these runs the model was handed over to the scheduling and fleet-planning branch for routine use for planning future fleet operations. The model makes it possible to check on a routine basis if scheduling is too tight, or too slack, in the sense of having too little or too much spare aircraft-time built in to cover late arrivals. Used thus, the model has proved to be a valuable aid to fleet planning.

Chapter 10

Capacity of an air traffic network

The capacity of an air traffic network is partly determined by the number of runways available at airports, which governs the rate at which aircraft can enter and leave the network at different points, and partly by the control rules used by the air traffic controller, which govern the rate at which aircraft can proceed along the airlanes between airports.

This study was concerned with how to increase the flow along the airlanes by changing the control rules to take advantage of the fact that a proportion of the aircraft possessed superior navigational equipment, and could therefore be relied upon to keep more accurately to their assigned tracks and speeds than others. It was proposed in fact to set up special tracks for the exclusive use of "equipped" aircraft, parallel to and alongside the existing tracks. These special tracks, although involving some extra mileage, would give equipped aircraft the chance of "getting away" when other aircraft were baulked by the ordinary tracks being fully occupied.

A high flow-rate along the reserved track could be maintained without use of radar control, because the enhanced navigation capability would enable the aircraft to fly closer together. Thus there would be no extra demand on radar controllers, and no extra expense in radar equipment. The capacity of the network would have been increased at minimum expense to the air traffic control authorities. The main expense would be to the airlines, in terms of the airborne navigation equipment needed on each aircraft, and the extra mileage involved on some routes.

But what would be the exact magnitude of the benefits in terms of reduced delay? Would they justify the expense of implementing the proposal? To answer this question a simulation model was constructed to represent firstly the current method of operation, and secondly the proposed method, so that a comparison could be made.

An actual terminal area, consisting of seven airports and surrounding airspace, was chosen for the simulation study. A busy summer day was selected as the basic time-unit to be reproduced in the model.

Aircraft entry times

The entry times of aircraft into the network were known from experience to vary from day to day due to airline delays (studied in the previous case-study), air traffic control delays in other airspaces, and the effect of wind on flying speed. But the entry times were not randomly distributed throughout the 24 hours; rather there were marked peaks and troughs of activity that occurred at roughly the same time each day. Therefore it was necessary to build into the model a schedule of flight entry times, and apply a random "disturbance" to each to represent the inherent variability.

Delay-producing factors

As noted above, the two main limiting factors in the system which restrict the flow of aircraft and cause delays are, first, the capacity of the airports to handle take-offs and landings, and secondly the density with which aircraft can be packed into the airlanes themselves. Airport capacity is determined principally by the number of runways available, and by the "airfield separation" that is used by air traffic control, that is, the time-interval between successive uses of a given runway. Ground congestion on taxiways can, on occasion, also be a source of delays. Several large aircraft "holding" at the end of the runway for route clearance may baulk other aircraft otherwise able to take-off, by blocking the path to the runway.

Airlane capacity, the second limitation on traffic flow, is determined principally by the "separations" used by air traffic controllers when allowing aircraft to enter the airlanes. These separations are based on (1) the speed difference between the aircraft entering the lane and the aircraft immediately ahead, (2) the positions and speeds of any aircraft that are known to be crossing the airlane or are otherwise "in conflict" with the given aircraft, and (3) whether the flight is to be monitored by radar.

76

Viewed as a system of queues, an air traffic system is unusual in that it is not possible to identify the delay sources as separate serving facilities, such as ticket desks or ship berths. The real limiting factor is space, not facilities. The movement of a given aircraft is controlled, not by the state of availability of facilities, but by the positions and movements of other aircraft relative to itself, and the need to maintain safe separations at all times.

Thus we have a system of traffic units (i.e. aircraft) flowing through a network consisting not of servers or facilities but of nodal points where controllers have the possibility of holding the traffic units, according to the position and movement of other traffic units in the network.

The aircraft play the part of both an active element, seeking to advance through the network, and a passive element, baulking other aircraft from making such advances.

The model

The model consisted of a number of program blocks or "sectors", one for each phase of an aircraft's progress through the network, from entry point—either on the ground at one of the airports or in the air—through to exit point.

At the beginning of each simulated day a schedule of actual aircraft entry times was made up by taking the schedule of planned entry times and adding a sampled delay to each. Time was advanced from event to event, most events being tied to movements of aircraft by aircraft "t-cells". The following events were represented:

(1) An aircraft, prior to flight, requests permission from air traffic control to start engines and commence taxi-ing to the holding point of the runway. This sector of program was entered whenever the model clock coincided with the entry time of an aircraft at an airport. Permission was granted unless more than a certain number of aircraft were already taxi-ing at that airport (in which case a delay was imposed whose value varied with the actual number taxi-ing). A taxi-ing duration was found by sampling, and this, added to clock time, gave the arrival time at the holding point.

(2) An aircraft joins the queue, if any, at the holding point.

(3) An aircraft, at the head of the queue for its route, requests an "airways clearance" from air-traffic control, that is, a time at which flight can commence, taking into account the last aircraft to use the airway route. Depending on the destination, speed, and altitude requested by the aircraft, and the positions and velocities of other aircraft in the air at the time, a first permissible take-off time is worked out, using air-traffic control rules as applied currently or as envisaged under the new system.

(4) An aircraft with airways clearance, and no other cleared aircraft ahead in the queue, attempts to obtain the runway for take-off. This attempt is frustrated if the runway is being used by a previous aircraft, or by an aircraft landing.

(5) An aircraft takes off, and enters the airway system.

(6) An airborne aircraft enters the system from elsewhere.

Programming and testing

The model was evolved from a hand simulation of the system. This hand exercise was continued during the programming of the model for a computer, so that some results would be forthcoming even if the program was not ready in time. In the event the program was ready, but the hand exercise proved an invaluable means of checking the computer program.

The computer used was an IBM 7090, and the program was written in CSL.

Use of the model

Validation runs were made by retrospective simulation of flights that took place in the previous summer. Delays to aircraft were then compared with statistical summaries of delays kept by an airline, covering the same period.

At first the two sets of figures were difficult to reconcile, but then it was realised that delays were being measured from different datum times in the two systems, so that like was not being compared with like. When this discrepancy was corrected, the two sets of figures agreed quite well. It was felt, therefore, that no significant effects or constraints had been omitted, and that the treatment was a valid one.

A series of runs were now made with different levels of traffic to compare the effect of introducing the new track into the network. It was shown that (1) the present system would become intolerably congested within a few years unless some improvement was made to the facilities available, and (2) the introduction and use of a reserved track within one of the airways would reduce delays to aircraft using it by about 20 per cent, the greatest effect being felt on the serious delays of over 30 minutes.

The results of this study served to quantify the advantages of exploiting this form of airborne navigational capability in congested terminal areas, and a number of operational trials are now in progress in several parts of the world to evaluate practical methods of putting this kind of technique into effect.

Chapter 11

Handling operations at a container-ship berth

Many handling schemes have been suggested for the loading and unloading of container ships—that is, ships specially designed to carry containerized freight. It was felt necessary to make a comprehensive study of the various possibilities, so that guide-lines could be developed to assist individual port authorities in choosing the most suitable scheme for their purposes.

One of the most important factors affecting the economics of container-ship operations is the utilization of ships. This is largely determined by the amount of time the ships spend in port between jouneys. When designing port facilities, therefore, it is necessary to ensure that quayside handling operations can be carried out as smoothly and as rapidly as possible, with the minimum of interference between different items of handling plant. In this connection it is useful if we can follow out (on paper or with a computer) the operations that have to take place on the quay, in advance of implementing any plan, so that the consequences can be viewed and improvements suggested. In the case now to be described, after attempts to do this by paper and pencil had been defeated by the magnitude, and tedium, of the task, it was decided to embark on a computer model. The handling scheme to be modelled was as follows.

The containers for loading (export containers) are arranged in pre-set order on a quayside park. When the ship arrives, the first operation is to unload the deck containers and convey them to spaces reserved for them in the park. Then the hold hatch covers are removed, and the next operation is to unload the containers in the hold. These are also conveyed to spaces in the park. At the same time the export containers referred to above are taken from the park and conveyed to the ship for loading into vacant spaces in the hold. In this way loading and unloading take place simultaneously, compressing the turnround time and making full use of the handling machinery. The last operation, after replacing hatch covers, is to load the export deck containers. The general scheme is shown in Fig. 11.1.

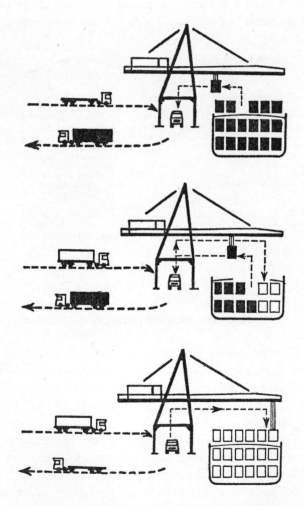

Fig. 11.1   Three handling cycles in container-ship turnround

There are two resources or items of equipment:

(1) quayside gantry cranes of the transporter type, and
(2) carriers (tractor-towed trailers).

Each of these follows a cycle of operations. The container lifting gear of the crane moves back and forth between the hold of the ship and the quay. The carrier moves back and forth between the quay and the container park. The way in which these two cycles interlock is shown in Fig. 11.2. There are two co-operative activities

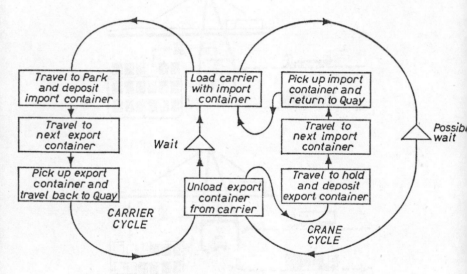

Fig. 11.2   Container-ship turnround model

which take place when containers are being exchanged between carrier and crane, the export container being loaded into the ship, and the import container being placed on the carrier. The other activities are simple operations involving only one or other of the items of equipment.

The moments at which partners arrive for their co-operative activities will not, in general, be synchronized because the "fetch-and-carry" times vary throughout the operation (the distances involved for each transport task vary). So one or other of the co-operating machines will be forced to wait.

We can, of course, reduce the waiting of one class of machine, say cranes, by increasing the number of partners available (carriers). But the waiting time of the carriers will then be increased.

In general, in these types of co-operative handling situations, it is true to say that the waiting, or loss of utilization, of resource A can only be reduced at the expense of the waiting, or loss of utilization of B, unless it is possible in some way, by re-arranging the sequence of tasks, to alter the sequence of cycle times so that the operations of the partners are better synchronized. We shall meet with an example of such re-arrangements below.

One of the objects of this present study was in fact to see how many carriers would be required to keep the quayside cranes busy without too many periods of waiting. Our approach was to construct the simulation model, then perform runs with different numbers of carriers available, and study (a) the overall turnround time, (b) the waiting time of the crane, and (c) the waiting time of the carriers. The queue length was not important because with the numbers of carriers we were using (2, 3 and 4), congestion on the quay would never hamper operations.

The model

The model was divided into three parts, performed successively, corresponding to the three phases of the turnround operation, namely:

(1) unloading deck cargo—a one-way operation involving import containers only;

(2) unloading and loading the hold—a two-way operation involving both import and export containers; and

(3) loading deck cargo—a one-way operation involving export containers only.

Only one crane was considered in the model.

In the first place the model consisted essentially of one event only: the co-operation of the crane and a carrier at the quayside, to place the import container on the carrier. On completion of this event, a "crane cycle-time" was allocated to the crane to determine its next available time (that is, the next time the crane trolley returns with a

G

new import container), and a carrier cycle-time likewise allocated to the carrier to determine its next available time (i.e. time of return to the quay for a new load). These cycle-times were read in by the computer as data. They were obtained by measuring the travelling distances taken from layout diagrams of the berth and ships, and using design speeds and accelerations for the equipment.

On completion of the first phase of the turnround, the next, two-way, phase was commenced in the simulation. This was represented in a similar fashion, with two co-operative events instead of one. The third phase followed which again consisted of one co-operative event.

Time was advanced in fixed six-second time-steps for the sake of simplicity.

The program was written in Fortran for a CDC 3200 computer, but it was later converted to run on a Univac 1108 machine.

Results

At the onset of the study it was proposed to arrange the park in three sections, corresponding to the three successive phases of the operations (see Fig. 11.3), as follows.

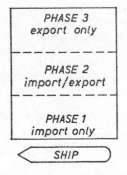

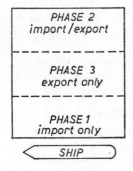

Fig. 11.3   Original layout of quayside park

Fig. 11.4   Final layout of quayside park

The first part of the park, nearest the ship, would be kept empty initially, so that it would receive imports during phase (1). The second part of the park, in the middle, would be reserved for containers of phase (2). The last part, furthest from the ship, would be kept for the export containers of phase (3). During the course of programming and data collection, however, it was seen that this was not the best arrangement because it would mean that the longest carrier cycles would be associated with phase (3), while the longest crane cycles would be associated with phase (2). In phase (2) the carriers could be delayed waiting for the crane, whilst in phase (3) the crane could be delayed waiting for the carriers. By simply re-arranging the park (see Fig. 11.4) so that phase (3) containers were allocated the space immediately behind phase (1) containers, a better balance in cycle-times could be obtained, with consequent shortening of the overall duration of turnround.

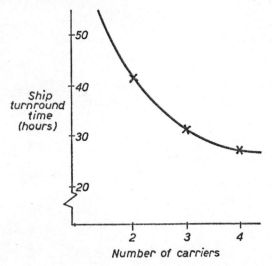

Fig. 11.5 Relationship between number of carriers and ship turnround time

The disadvantage of such a rearrangement would be that control over the movements of the carriers would be made more difficult. However, it was decided that the gain in terms of reduced turnround time would outweigh this disadvantage.

When data specifying the travel times relating to this new park layout had been punched into cards ready for computer input, runs were made with the model to study the relationship between number of carriers and ship turnround time. The crane working speeds were varied to see what effect different assumptions would have on the result. Fig. 11.5 shows a typical example of the relationships we obtained. As with the standby aircraft of Chapter 9, we found a law of diminishing returns operating, due to the fact that opportunities for delay-saving arise less often as the fleet size increases.

These results were passed to individual port authorities so that they could assess their own needs in regard to carriers on their berths, taking into account on one hand the cost of carriers and on the other the need to achieve rapid, delay-free ship turnrounds.

Chapter 12

Plant utilization and storage

In this chapter we describe a study of activities in the Finishing Department of a tubeworks. The input to this department consisted of tube, newly produced in an adjoining shop by hot-rolling from continuous steel strip. The task of the Finishing Department, laid out as shown in Fig. 12.1 (overleaf), was to provide any final treatments required by customers, such as galvanizing, varnishing, screw-and-socketting, or hydraulic testing.

Due to the wide variety of possible finishing treatments, numerous machines were needed in the Finishing Department, and these were dispersed over a wide area of floor. Handling was by overhead crane. Material flows are shown in Fig. 12.2.

Buffer stocks

It was necessary to hold buffer stocks in the Finishing Department to enable finishing work to proceed at a steady rate without being held up for lack of suitable tubes to process. These buffer stocks were stored at several points (as shown in Fig. 12.1) situated within the finishing area and elsewhere. Floor space was short and the management wished to know (1) whether the storage requirement could be reduced by re-organizing the operations of the Finishing Department, (2) whether planned new finishing machinery would imply extra, and unacceptable, storage requirements.

Studies of improvements

In their attempts to obtain improvements, Production Management recognised the limitations of existing finishing arrangements. They therefore called for a Work Study survey. The initial survey indicated the magnitude and complexity of a detailed investigation, and simulation techniques were considered. Operational Research

87

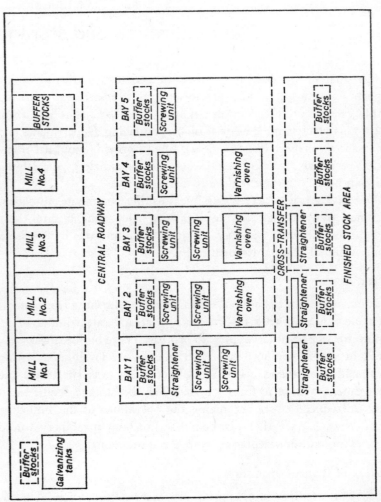

Fig. 12.1   Layout of plant in a steel tube works

staff were called in and took the view that a computer simulation would be appropriate. A number of possible ways of improvement were foreseen, but the main effects to be evaluated were those on stock levels. Simulation seemed an obvious tool for trying out a wide variety of policies and circumstances. Subsequent investigations and study work were therefore planned with a view to simulation, and eventually a model was formulated and programmed for the computer. For simplicity, it was decided to exclude Bays 4 and 5.

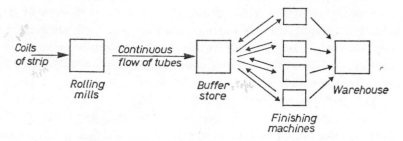

Fig. 12.2   Flow of material in tube works

Basic unit of material

As explained above, overhead cranes were used for handling. Crane "hoists" varied from 3 to $4\frac{1}{2}$ tons of tube, depending on tube size. Tubes progressed through the Finishing Department in "hoist" bundles. It was possible therefore to use hoist bundles (or "hoists") as the units of the simulation, thus reducing the number of units to manageable proportions.

Production schedule

Each week a schedule was made up for the following week's production. Incoming orders from customers were batched by tube specification and finish, the batches being then arranged in a time sequence for production at the rolling mills. This sequence was chosen to minimize the number of rolling-mills size changes, and to spread the work-load on finishing machines as evenly as possible over the week, so as to minimize peaks of buffer stock-holding.

It was decided to take this weekly production schedule in the form of a list of batches as the basic input to the model. The arrival of discrete units of material could then be generated as the run progressed, by working through each batch in turn, using inter-arrival intervals calculated from mill rolling rates to compute the time of arrival for each new hoist.

As each unit arrived it would either be sent to a finishing machine, if a suitable one was available, or put into buffer.

Work sequencing

One of the most difficult parts of the study was to discover, and formulate, the rules used to control and sequence the work of the Finishing Department. By putting many hypothetical situations to shop foremen, certain broad principles based on common sense emerged, and from these it was possible to construct rules suitable for use in the model.

The first obvious point was that tube-bundles, being stored one on top of another, tend to be removed from storage in the reverse order to that in which they are inserted. "Digging" is time-consuming and thus discouraged. This led to the adoption of a LIFO (last in, first out) rule of selection, in the model.

The second point was that tubes of different diameter often require different machine settings. To avoid loss of machine utilization, therefore, through frequent time-consuming re-sets, size changes are only made if no other material can be found in buffer for the machine in question.

A third point was that handling operations can be reduced if material can be routed direct from rolling mill to finishing machine, and direct from one finishing machine to another (where two treatments are specified), without having to be placed in buffer, and drawn out again later. From the point of view of the model this meant that when deciding on the next tube-bundle to process on a given machine, those newly arrived from the mill, or having been worked on at other machines, should be given priority.

Handling

There was spare handling capacity in the part of the plant simulated; crane utilization was less than 50 per cent. It was decided, therefore, in the interests of simplicity to omit cranes from the model.

This amounted to making the assumption that in any proposed change to plant operation crane capacity would not act as a limiting factor. It was felt that this was a reasonable assumption to make initially. Later, if necessary, cranes could be included.

The model

The model that was finally formulated consisted of the following six basic event types:

(1) *Arrival of hoists.* Using the production schedule and inter-arrival intervals derived from tube-mill rolling rates, a stream of arriving hoists was generated. On arrival, these hoists became available for allocation to a finishing machine, or buffer storage.

(2) *Finishing operation.* When a hoist became due for allocation (either having just arrived, or else having just finished at a previous machine), a suitable machine was sought to take the hoist and (if one was found) a finishing operation was then simulated by adding a sampled duration for the finishing operations to the time of availability of both the machine and the hoist. Whenever a machine became available, a suitable hoist was sought in buffer.

(3) *Hoist into finished stock.* If a hoist had received all required treatments, it was sent to finished stock. Statistics relating to plant throughput were then updated.

(4) *Hoist into buffer.* If after considering all possible events of types (1) to (3) no destination could be found for a hoist, it was put into buffer. Statistics relating to buffer size and composition were then updated.

(5) *Machine set-up.* If after considering all possible events of types (1) to (3) no suitable hoists could be found for a machine, a resetting of the machine to a different size was considered. Buffer stocks were scanned, and the number of hoists there of each size-class counted. The size-class having the maximum hoists was then selected as the new size for the machine.

(6) *Shift changeovers and tea breaks.* Some machines run down at the end of a shift. This was represented in the model by "flagging" these machines as unavailable for a short period at shift end.

Run-up at the beginning of a new shift was represented in a similar way. The all-important mid-shift tea break was also included in the model.

The program was written initially for an IBM 7094/II computer using the CSL I language; later it was transferred to run on a CDC 3200 machine. Running speed on the latter machine was about 1·5 weeks per hour.

Output information consisted of a shift-by-shift printout of the size and composition of buffer stocks, a cumulative total of the number of hoists processed by size and composition by each machine, the utilization time of each machine, and the average throughput times of different classes of tube.

Unexpected varnishing load

After the programming and logical faults in the computer program had been ironed out, we attempted to simulate the first week of a chosen four-week period. The computer got only as far as the fourth day before it calculated that the queue for Varnishing was exceeding the bounds of storage set aside for it. These bounds had been agreed in advance and represented a very generous maximum queue size which it was thought would rarely be approached in practice.

Clearly something was amiss. We scrutinized our rules and could find no fault with them. The answer was provided by the plant. The load on Varnishing in Bays 2 and 3 was indeed heavy, but this fact was obscured by the practice of storing the pre-varnishing stocks in the warehouse, which was not part of the area under study. Furthermore it was a regular practice to work a percentage of the varnishing load arising in Bays 1 to 3 on the varnisher of Bay 4, also not being studied, and to do the remainder of unfinished material at the weekend on a special Sunday shift.

This illustrates the danger of studying one part of a plant in isolation. It also exemplifies a very common experience in Operational Research work, namely that the problems are often not where management believe them to be. We had been asked primarily to study the pre-testing/screwing/straightening stocks. We found at this early point in the study that more serious queues existed for varnishing, and if anything it was varnishing capacity that was short, not testing/screwing capacity.

Final validation runs

To correct the varnishing situation we posited two additional rules. First, 15 per cent of all arrivals for varnishing in Bay 3 would be transferred across to Bay 4, or, in terms of the model for Bays 1 to 3, 15 per cent of these arrivals would be deleted from the system at this point. Secondly, at the end of shift 16, the Saturday morning shift, any excess over 25 hoists would be assumed to be worked on a Sunday shift and would thus be completed by the beginning of shift 1 of the following week.

With these new rules we ran the model again, with results as shown in Fig. 12.3, from which it can be seen how the buffer builds up very rapidly over the first few shifts, then seems to stabilize at about 100 units, and has a final spurt in the last two shifts of the week, reaching a grand total of 130 at the finish. Clearly the two varnishing units in Bays 2 and 3 were quite unable to cope with the demand during this first simulated week. The broken line shows the cumulative transfer to Bay 4, which becomes quite significant taken over the week as a whole.

Fig. 12.4 illustrates the behaviour of the collective buffer of material awaiting testing, screwing, or straightening. This buffer shows a much more stable behaviour, indicating that the testers, screwers, and straighteners are able to cope with the demands made on them. It is interesting to note that the queue shows no simple trend. At first it lengthens, then it seems to be dying away, but recovers at about shift 12 and remains relatively stable thereafter. This is due to the irregular nature of the arrivals.

These results turned out to be close to the actual experience of the plant during the period we were simulating. The model could therefore be said to be validated.

Results

At the time of the study the screwing/testing machines were being manned by crews of eight men. By suitably modifying the machines it was possible to arrange for the testing-only operation to be performed with a reduced crew of only four men. Management therefore proposed to keep an eight-man crew working together on one machine when screwing, but to split into two four-man crews working on two machines when testing. The advantage of this method

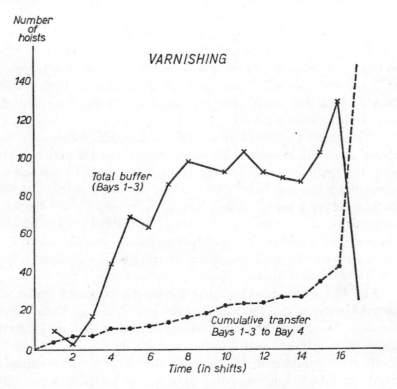

Fig. 12.3 Queue for varnishing of tubes

was that it enabled material to be tested at twice the rate by the same number of men. No extra machines were needed because not all machines were being manned under the current method of working.

The proposal actually went a stage further. If the same number of men could perform more work under the proposed scheme, it followed that fewer men were needed to perform the same work. At the time, four complete eight-man crews were employed. Could three such crews manage on their own if they adopted the new method of working?

The work-load on the machines was about equally divided between screwing and testing work. From the point of view of machining hours available and machining rates, three crews should therefore be able to provide sufficient capacity in the plant to complete all the work currently being undertaken.

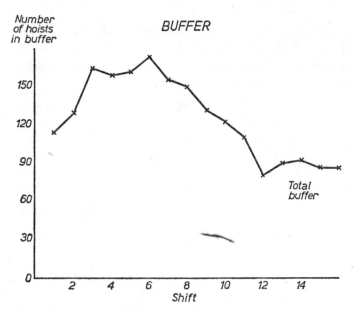

Fig. 12.4   Queue for test, screw, or straighten

Management did not know how the buffer stock position would be affected by this plan. It was possible that additional bottlenecks might be created which would add to the congestion, increase the peaks of buffer stock-holding, and thus make the scheme unattractive or even unworkable.

Runs with the model proved conclusively that such would not be the case. Buffer stocks would increase, but would never become excessive. Additional runs showed that by augmenting the work force by a special four-man testing-only crew, extra material could be processed which was being currently sent out to be finished in another plant some distance away.

It was this final scheme that was finally put forward to the tube works management. Its advantages were seen to be:

(1) reduction of labour by four men per shift—a saving of twelve men;
(2) increased production;
(3) saving the cost of shipment to the remote factory.

Temporary modifications were made to the machines to prove the principle in practice. These were successful, but indicated that more extensive modifications would be required before permanent introduction of the scheme would be possible.

Chapter 13

Comments on model-building

This chapter is devoted to a general discussion of the practical aspects of model-building. Much of the material is based on the case studies of the previous four chapters. But the discussion applies equally well to the econometric, military, and other models described in earlier chapters. ⌐7

Problem definition

For any model-building activity, the first requirement is to have a very clear statement of the problem to be solved. Otherwise it is impossible to make the many simplifying assumptions that are needed in order to formulate a manageable model. For instance, in the case study described in Chapter 9, many of the factors governing aircraft allocations in the real-life situation were judged irrelevant to the model on the basis that they were seldom, if ever, allowed to interfere with the punctual running of flights. If the aim of the model, namely to estimate the frequency of flight delays implied by alternative operating plans, had not been clearly understood, many of the omitted factors might well have been included, complicating the structure of the model without adding to its problem-solving power. Indeed, model outputs would have been more difficult to interpret.

Unfortunately it is often the case that a simulation is proposed without anyone having a clear idea of what it is expected to achieve. There is a vague feeling that once a model is available enabling certain real-life operations to be simulated, all sorts of questions can be answered. This, of course, is fallacious because a model can only answer those questions that it has been tailored to answer. As a result, when the model has been constructed, often at considerable time and expense, it proves to be a source of disappointment because of its inability to answer questions posed, without under-going extensive modification.

There is one case in which it may be justifiable to commence model-building without a clear idea of the objectives, namely when this seems to be the only way of defining the objective. By producing some sort of model it is possible to collect reasons why the model is inappropriate! Thence it may be possible to deduce what the real problem is, and what form a more appropriate model would take. However, this is a rather dangerous procedure inasmuch as the model-builder is likely to find himself associated with the first model thereafter.

Studying the real-life system

When the objectives of the study have been defined, the next stage is to study the current operation of the real-life system so as to see what factors currently affect it, which of these can be modified, and what the scope for improved operation is likely to be. This is best done by interviewing those closest in touch with the operation, namely the supervisors and line managers. These interviews should be backed up with some detailed perusal of written records, so that opinions can be tested against unbiassed facts, and some idea of magnitudes gained. Whether some factor is said to be important or not may depend very much on who is asked. Records can be used to throw light on the number of times such a factor had an effect. If shown to be infrequent in the period of time over which the model is to be run, the factor can usually be ignored, in spite of the opinion of the operating staff to whom it may well seem important.

Having considered current factors, it is then necessary to inquire what changes are likely to affect the system in the future. In some cases the whole operation is a hypothetical one, not existing at the time of the study; the case study dealt with in Chapter 11 was an example of this. Then the factors that will be important can only be intelligently guessed at. Some guidance can be obtained from the manufacturers of the equipment to be used, but their claims for the performance, reliability and so on of their product must be treated with a certain amount of caution, and in some cases, downright scepticism. The model-builder may be able to base some of his assumptions on the experience of people who have worked with similar systems, but for the rest he may have to rely on common-sense.

Preliminary model

When the significant factors have been isolated, and the working of the system is reasonably well understood (and not before) then it is time to start constructing a model of the system. This model should be regarded as a preliminary model, not a final version. The object should be to produce a model which can simulate known conditions, so that the validity of the approach can be tested. As a result of such validation runs, some modifications will almost certainly be needed to bring model performance into line with the actual performance. These modifications may involve simple changes to parameter values. It is more likely, however, that they will involve changes in the logical structure of the model in certain areas. This likelihood should be borne in mind from the start so as to avoid disappointment with the first model, which should properly be regarded as only a preliminary stage in the model-building process.

Before attempting to put the model onto a computer, it is a good idea to try out the simulation on a small scale, by hand. Many logical errors can be removed at this stage, quickly and cheaply, and much time in the programming thereby saved. The rules of the simulation should be written down in the form of a flow chart. Then attempts should be made to play out part of a simulation according to these rules.

At about this point in the study the model-builder will have to decide on the repetition period to be used in the model. There is nearly always some natural period in the system which settles this question. For instance, in the air traffic simulation of Chapter 10, a 24-hour period was a natural repetition period, although, had day-to-day variations in traffic been found to be important, a week would have been the appropriate period. In the case of the tube works simulation of Chapter 12, the basic repetition period was the four-week production cycle. Actual runs of the simulation model will, of course, cover longer periods because, in general, several

H

repeats of the basic period will be needed in order to average out chance sampling fluctuations (see Chapter 8).

Having fixed the length of the basic period, the model-builder can then select an actual period of time to use as a basis for validating the model. He can then make a start on collecting data about the performance of the real-life system during this period. It is a good idea to start this data-collection task as soon as possible, because it can be time-consuming and could hold up the progress of later stages of the study.

Programming

No attempt should be made to program the model for a computer until the above stages have been carried out. Programming is a complicated and costly activity, not to be entered into lightly. To work efficiently, the programmer needs to have a complete specification of the program before he starts coding. He will often find it difficult to accommodate changes once he has started work. It is therefore most important to do as much thinking and research as possible before handing the model over for programming. This will reduce, if not eliminate, changes called for during and after initial programming, and shorten the duration of the study as a whole.

For the actual programming the use of a simulation programming language is worth considering. Some information about these languages is given in the Appendix. Technically, there is much to be said for the use of one of the languages, since in this way considerable programming time can be saved and the job of making program changes can be simplified. But these benefits only accrue if the programmer is thoroughly familiar with the simulation language he is using. If, as too often seems to be the case, the programmer has not tackled a simulation model before, he will have to spend some time finding out about the little points of detail which appear in no manual but which nevertheless determine whether the program functions correctly or not when run on the computer. Another consideration worth bearing in mind is that the program may well have to be handed over to another programmer at a later stage. If he, too, is unfamiliar with the simulation language, another learning period is involved. The conclusion is that unless there is a pool of programmers engaged almost exclusively on simulation work, it is better to stick to a general-purpose language such as Fortran or Algol and pay the extra programming time penalty that this entails.

The problem of estimating programming time is bound to arise. It is very difficult to foresee how long the process of developing a simulation programme will take because of its complex logical structure. Experience shows that programming time depends more on logical complexity than on the number of instructions or any other measure of program size. A program with few branches is relatively easy (i) to conceive correctly in the first instance, and (ii) to check when written. Conversely a program with many branches is (i) very liable to contain logical errors due to faulty conception, and (ii) likely to be difficult to check because some branches will be infrequently entered during program execution. A good index of programming time is therefore the number of branch instructions in the program. It is usually possible to judge roughly the number of branch instructions even before the program has been written, by considering the logic of the model, especially if this has already been expressed in flow chart form. Then if a standard number of programming hours is associated with each branch instruction, based on past experience, a fair estimate of the total programming time can be obtained. This type of calculation can be very salutary if done at the beginning of the programming phase of the project.

Whether this particular method of estimating is adopted or not, it is important to use some objective measures of programming performance when trying to foresee how long the programming task will take. Subjective estimating based on "experience", in other words on hunch, is liable to be too optimistic when faced with programs of considerable complexity. Surprise is then registered when deadlines are not met. Surprise may turn into frustration and even disillusion as the completion date is repeatedly put back.

Estimates of model running speed will also be called for at some stage prior to completion of the program. Again it is important to have some objective measures. The best method here is to consider the number of events that are required to be simulated in the model for each unit of real time, and then to estimate how long an event will take on the computer, from data on computing speeds of the machine in question. In the air-traffic model, for instance, we asked how many aircraft movements would have to be simulated in the model to represent one hour of real-life operation of the air terminal. Then we asked how many events, on average, would have to be simulated for each aircraft movement. With the answer to these questions we were able to see how long the computer would take to simulate any given period of real time.

Validation

The process of validation is straightforward enough in principle: the model is made to simulate the behaviour of the system under the exact conditions that prevailed in some past period, and the results are then compared with what actually happened during that period. But, in practice, conclusive validation tests are difficult to arrange. The first difficulty arises from the fact that the model outputs are in general probabilistic in nature. We have seen in Chapter 8 how the variability of model outputs can be reduced by increasing the length of simulation run. It is unlikely, however, that the validation period will itself be very long, due to the labour involved in collecting data. Over such a period, model outputs may well be quite variable, and any comparison between actual and model is thus difficult to interpret. The model will have to be run several times with different random number seeds to indicate the range of variation. If valid, these model results should bracket the actual results. But even if they do so, there may still be consistent bias in the model which is concealed by the small sample of model results available. It is thus difficult to draw convincing conclusions from these comparisons.

Another practical difficulty associated with validation is that it is hard to find a suitable period of time to use. Ideally the period should be as long as possible, for reasons just explained; but it is also important that factors affecting system performance should stay as constant as possible throughout the period. These two requirements are difficult to reconcile. It is also desirable to exclude any special events, such as bank holidays, which might disturb the results and invalidate comparisons based on average conditions.

Another point is that it is often very difficult to obtain the necessary data on the real-life system. For instance, in the tube-works simulation described in Chapter 12, it proved to be difficult to find out exactly what the stock position was at any point in time. The tube bundles were often inaccessible, and the state of the various bins was continually changing as production proceeded.

In spite of the foregoing, validation should always be attempted, because, however inconclusive, it does provide a check against the grosser errors, and gives the model-builder confidence to go on and use the model for prediction.

Use of the model

Before the model can be used, the problem of run length has to be resolved. This, as we saw in Chapter 8, depends on (i) the variability of the outputs, and (ii) the cost of computing time. During preliminary test runs it is possible to determine both these factors. It is likely that the computing cost will turn out to be such as to rule out the use of statistically adequate runs. This is the unfortunate experience of practitioners: they are never able to do as much simulation as they would wish.

However, even with limited runs, much information can be gained about the performance of the system. Compared with what is possible by hand, computer simulation can deal with very large samples indeed. Even if some margin of statistical error has to be assigned to the results, due to limitations on run length, the results are nevertheless much better than any that could be obtained by hand.

Very little can be said about the actual experiments that should be performed once the model has been completed. Systematic programs of experiments in which factors are varied one at a time, or according to some statistically designed scheme, are rarely undertaken in practice. The model-builder usually knows too much about the structure of his model to adopt such a blind approach. The usual procedure is to analyse the results of each run, and try to see where the model system is most in need of improvement. Further runs can then be performed to test the effect of the improvements made in these areas.

This approach has the virtue that each run gives a little more insight into the working of the model, and hence into the working of the real-life system which is being represented. Nearly always it is this enhanced understanding of the system which is the real pay-off from simulation studies.

Appendix

Programming languages

This appendix is devoted to a discussion of programming aids currently available for models of the type described in Chapter 6. The computer programs required to implement these models possess some special features not needed in programs designed to solve other types of problems. The simulation programmer therefore needs something more than the facilities provided by a scientific high-level language such as Fortran or Algol. Two main approaches have been adopted to the problem of providing extra facilities, namely (1) *the program-package approach*, which consists of providing a set of useful program routines which can be called on by the simulation programmer; and (2) *the language-compiler approach* which consists of creating a special simulation programming language, together with a compiler program to translate it either to machine code or, more commonly, to one of the already established programming languages.

The first approach has several points in its favour. Firstly the package, if written in one of the standard programming languages, namely Fortran or Algol, can be compiled and used on any machine with a compiler for that language. In practice this makes the packages very widely available.

Secondly, anyone already versed in the language adopted can understand and use the simulation package with a minimum of instruction. Also, if, as so often happens, the simulation programmer has to divide his time between simulation work and other scientific computing work, the fact that he can use effectively the same language for both has obvious advantages.

The disadvantage of the package approach is that the facilities that can be provided are somewhat limited. To obtain a fuller range of facilities a language-compiler approach must be resorted to.

105

The compiler approach on the other hand, while providing these superior facilities, suffers from the disadvantage that it can only be used on those machines having the appropriate compiler. Another point to bear in mind is that a period of several months is needed to gain full proficiency in a new programming language of this type.

In passing it is worth noting that the Rand Corporation are experimenting with a third approach, namely a questionnaire processor which creates a simulation program (in Simscript language) from a questionnaire input.

The attractiveness of this approach is that it eliminates the need to write a program, thus saving time and expense. On the other hand, there is considerable loss of flexibility due to the fact that only specified answers to the questions can be handled. The questionnaire, therefore, does not provide a truly general-purpose vehicle for simulation modelling. Not enough experience has been acquired to make an assessment of the potential of this approach, though it would seem to provide an excellent medium for the situations which it fits.

Facilities needed by the simulation programmer

We can list four programming facilities commonly needed by the simulation programmer and which are not provided by general high-level programming languages. We shall name below some individual simulation languages and program packages that provide these.

(1) *Automatic time advance.* We have seen in Chapter 6 the need for some means of advancing the model clock from event moment to event moment and ensuring that all current events are simulated before the clock is moved on. It is clearly an advantage if a facility is provided to advance time automatically, once all relevant events have been processed.

(2) *Facilities for specifying decision rules.* Complex decision rules form an essential part of most models. A decision rule can be expressed as a rule of choice among a set of alternatives. (The simplest example is the rule governing the choice of next customer to serve when several are waiting.) It is convenient, therefore, to have some type of list structure available, so that such rules can be specified

in terms of "take first list member", "take last list member", or "take any list member with attribute X".

(3) *Random sampling facilities.* Nearly all simulation models involve taking random samples from specified probability distributions. We have seen in Chapter 2 that the basic requirement here is a source of pseudo-random numbers. In other words the simulation language (or package) should have a pseudo-random number generator, suited to the machine on which the model is going to be run. This is usually provided as a subroutine, coded in machine language (for speed of execution), the actual method of generation being based on the multiplicative congruence method explained in Chapter 2.

In Chapter 8 we explained the usefulness of the technique of having different random-number streams for different random variables in the model, to enable the closest possible comparisons to be made when experimenting with changes in the controllable factors. This technique is possible if the programmer can specify in his program the starting number for the random-number generator. The same generator can then be made to produce different streams by specifying different starting numbers for each separate random variable in the model.

In addition to a generator, the programmer commonly requires exponential and normal deviates, and samples drawn from other standard probability distributions.

Also it is useful to have facilities for inputting a histogram and sampling from it, to enable empirical data on random fluctuations to be used where theoretical distributions are inappropriate.

(4) *Facilities for analysing results of runs and producing reports.* The results of simulation runs are normally required in the form of statistical summaries of the mean values and variations of model variables. For instance, in a queueing problem the mean waiting time and the distribution of waiting times are required, often broken down by time of day, type of customer, and other factors.

It is useful, therefore, to have a means of easily specifying the variables to be recorded, the distributions required (what intervals, etc.), and the layout of the results as finally printed, along with titles and other explanatory text.

Such facilities can reduce programming effort enormously. The writing of output routines is a well-known consumer of programming time!

Following the pioneering work of K. D. Tocher at United Steel, many simulation languages or program packages for simulation are now available. A list of those known to the author is given below. All provide the facilities discussed above. For further information on individual languages or packages the reader is referred to descriptions and manuals obtainable from computer manufacturers or other sponsors, as shown below.

(1) Languages using the Compiler approach

Language	Machines	Authors	Sponsors
CSL	IBM 7090 Honeywell 200, 400 ICT 1900 EE KDF 9	Buxton Laski	IBM–UK and Esso (U.K)
GPSS	IBM 7090 IBM 7040 IBM 360 Univac 1107/8	Gordon	IBM USA (more used)
Simpac	IBM 7090	Lackner	System Development Corp.
Simscript	IBM 7090 IBM 7040 CDC 3600, 6600	Markowitz Hausner Karr	Rand Corp. (U.S.A)
Simula	Univac 1107/8	Nygard, Dahl	Norwegian Computing Centre USA

(2) Program Packages

Name	Machines*	Authors	Sponsors
E.S.P. (Algol)	Elliott 503, 803	Williams	Elliott
Gasp (Fortran)	IBM 1620, 7070, 7090	Kiviat	U.S. Steel
Simon (Algol)	Elliott 503, 803 EE KDF 9, ICL Atlas	Hills	Bristol College, U.K.
Sol (Algol)	Burroughs B5000	Knuth, McNeley	Calif. Inst. Tech., and Burroughs

*Machines on which the program is known to have been run. There should be no great problem in running on other machines possessing suitable compilers.

References

[1] H. S. KRASNOW, "Dynamic Representation in Discrete Interaction Simulation Languages", in *Digital Simulation in Operational Research* by S. H. Hollingdale, English Universities Press, 1967.

[2] T. H. NAYLOR et al., *Computer Simulation Techniques*, Wiley, 1966.

[3] K. D. TOCHER, "A Review of Simulation Languages", *Op. Res. Qly* **16**, 2, June 1965.

Index